The Universe Next Door

The Universe Next Door

The Making of Tomorrow's Science

Marcus Chown

OXFORD
UNIVERSITY PRESS
2002

OXFORD
UNIVERSITY PRESS

Oxford New York
Athens Auckland Bangkok Bogotá
Buenos Aires Cape Town Chennai Dar es Salaam Delhi
Florence Hong Kong Istanbul Karachi Kolkata
Kuala Lumpur Madrid Melbourne
Mexico City Mumbai Nairobi Paris São Paulo Shanghai
Singapore Taipei Tokyo Toronto Warsaw

and associated companies in
Berlin Ibadan

Copyright © 2002 by Marcus Chown

Published by Oxford University Press, Inc.
198 Madison Avenue, New York, New York 10016

Oxford is a registered trademark of Oxford University Press

Library of Congress Cataloging-in-Publication Data
Chown, Marcus
The universe next door : the making
of tomorrow's science / Marcus Chown.
p.   cm.   Includes index.
ISBN 0-19-514382-5
1. Physics—Popular works.
2. Cosmology—Popular works.
3. Life on other planets—Popular works.
I. Title.
QC24.5 .C47 2002   523.1—dc21   2001036400

1 3 5 7 9 8 6 4 2

Printed in the United States of America
on acid-free paper

To Karen (the jewel that is my wife!)
with love, Marcus

Contents

Part Three: Life and the Universe 121

listen: there's a hell
of a good universe next door; let's go

—E. E. Cummings

The Making of
Tomorrow's Science

Among all the services that can be rendered to science, the most
important is the injection of novel ideas.

—J. J. Thomson

What we need is imagination. We have to find a new view of the
world.

—Richard Feynman, *The Meaning of It All*

New ideas are the stuff of science. Without a constant supply of
them—clay pigeons catapulted into the blue sky to be shot down—
science would be impossible. As the cosmology consultant of *New Scientist*, I often come across ideas that blow my mind, that leave my head
spinning with their far-reaching ramifications. Such as the notion that
time could actually run backward; or that there might exist multiple realities playing out all possible histories; or that our universe may have
been created as an experiment by superior beings in another universe.

Invariably, such ideas are attempts to answer the big questions in science. What is time? What is reality? Are we alone in the cosmos?
Where did the universe come from? Like nothing else, these questions
expose the limits of our current knowledge, highlighting the key issues
with which scientists at the frontier are wrestling.

What follows are my dispatches from the frontier of the imagination.
At first sight, the ideas may seem crazy. But then, once upon a time,
the idea that time slows down for someone traveling fast or in the presence of gravity seemed crazy. Now, "time dilation" can be demonstrated with superaccurate atomic clocks and nobody seriously doubts
it. Once upon a time the idea that an atom could be in two places at
once—the everyday equivalent of being able to sit down and stand up

at the same time—seemed crazy. Now, not only is this easily demonstrable, but inventions that exploit the ideas of quantum theory are estimated to account for 30 percent of the gross domestic product of the United States.

"Craziness," therefore, is not necessarily a ground for dismissing an idea. Nature is under no obligation whatsoever to respect our sensitivities and behave in a way that appeals to everyday common sense. "Your idea is crazy," the great physicist Niels Bohr is reported to have told a colleague. "The question is: is it crazy enough to be true?"

Of course, the scientific imagination must work within the limits of the known facts. And there is evidence for all the ideas presented here. This book is a tribute to extraordinary people with extraordinary ideas. It's a salute to those with the courage and imagination to try to make tomorrow's science. It's an homage to those who are struggling to see beyond the edge of the known universe.

I hope that in reading this book you will get some feeling of what a wonderful, weird universe we find ourselves in—a universe far stranger than anything we could possibly have invented. And I hope it gets you thinking. Without further ado, then, and in the words of e. e. cummings: "listen: there's a hell of a good universe next door; let's go."

Acknowledgments

My thanks to the following people who either helped me directly, inspired me, or simply encouraged me during the writing of this book: My dad, Karen, Sara Menguc, Lindsay Symons, Kirk Jensen, Jane Lincoln Taylor, Helen Mules, Gregory Chaitin, Patrick O'Halloran, Nick Mayhew-Smith, Sir Fred Hoyle, Allison Chown, John Cramer, Cliff Pickover, Sir Martin Rees, Michael Brooks, Stephen Battersby, Andy Hamilton, Elisabeth Geake, Alex Jones, Garry Williams, David Hough, Stephen Hedges, Sue O'Malley, Pam and Mike Young, Spencer Bright, Karen Gunnell, Pat and Brian Chilver, David and Pauline Parslow, Stella Barlow, Barbara Pell, Maureen Butler, and Juliet Walker. I would especially like to thank those I have interviewed: Max Tegmark, Larry Schulman, Edward Harrison, Humphrey Maris, Mark Hadley, Keith Dienes, Mike Hawkins, Robert Foot, Sergei Gninenko, David Stevenson, Chandra Wickramasinghe, and Alexey Arkhipov. (None of these people is responsible for any errors.)

Part One
The Nature of Reality

1

Unbreak My Heart

*Contrary to all expectations, there may exist regions
in our universe where time runs backward*

Eating is unattractive. . . . I select a soiled dish, collect some scraps
from the garbage, and settle down for a short wait. Various items
get gulped up into my mouth, and after skilful massage with
tongue and teeth I transfer them to the plate for additional sculp-
ture with knife and fork and spoon.

—Martin Amis, *Time's Arrow*

If you go flying back through time and you see somebody else flying forward into the future, it's probably best to avoid eye contact.

—Jack Handy

Out in space a star unexplodes. Moments ago, a blisteringly hot
shell of tortured matter was flying through the vacuum at thirty
thousand kilometers a second. Now the last shreds of glowing debris
have been sucked back into the star. Already, it is embarking on the
long journey back to the time when it will be unborn in a cold cloud of
interstellar gas.

Surely such a sequence of events is nonsense? Not according to a
prominent American physicist. In a scientific paper published in the
very last week of the twentieth century, Lawrence Schulman of New
York's Clarkson University dropped a bombshell into the world of
physics. He showed that there could exist regions in our universe
where time actually runs backward—where stars unexplode, eggs unbreak, and people grow younger with every passing second.

How can time possibly run backward? Well, first it is necessary to
understand why it runs forward.

3

The Arrow of Time

The normal direction of time is associated with things growing old and falling apart. Think of a photograph of a coffee mug and a photograph of the same coffee mug broken in pieces. Which picture was taken later? The photograph of the shattered mug. Without fail, everyone would associate the past with the intact mug and the future with the broken mug. Common experience tells us that mugs do not spontaneously unbreak. But why don't they?

The answer is not at all obvious. A striking feature of the fundamental laws that underpin the universe, including those that govern the atoms that are the building blocks of coffee mugs, is that they are almost exactly time-reversible. In other words, any process that occurs one way is also permitted to occur in reverse. For instance, an atom can spit out a photon of light and it can also suck in a photon of light. So if you were shown a movie of an atom doing something, you could never tell whether the movie was being run forward or backward. Both events would appear perfectly reasonable.

In light of this, the question of why coffee mugs do not unbreak can be put more precisely. Why is it that when particles such as atoms club together to make a coffee mug, a direction of time is imposed where previously none existed? The answer is subtle. Surprisingly, it has to do with the idea that there are far more ways in which a coffee mug can be broken than ways in which a coffee mug can be intact.

Think of all the possible ways a mug could be shattered. It could be shattered, for instance, into one big piece and ten small pieces, or two big pieces and sixteen small pieces, or two big pieces, twelve small pieces, and a lot of china dust. . . . You get the idea. Clearly, the mug can be broken in a huge number of ways. Okay, now think of all the ways that the mug could be intact. There is one, and only one, way. Consequently, if all possibilities are equally likely, it is extremely probable that a mug will go from being unbroken to being broken. There are simply many more ways it can be broken than ways it can be unbroken. It's not completely impossible that, when a collection of broken fragments is hit by a hammer, the fragments will leap back together into the form of an unbroken mug. It's just overwhelmingly unlikely. It would be necessary to wait many times longer than the current age of the universe to have any hope of seeing such a thing happen.

"It's easy to break a mug, difficult or impossible to put the pieces back together again," says Schulman. "The choice of the easy direction physicists call the arrow of time."*

The arrow of time is also the direction in which metal rusts, medieval castles crumble into ruins, and people grow old and die. What all these processes have in common is a change from a state of relative order to one of relative disorder. And each and every one of them occurs for the same reason: there are many more disordered states open to a body than there are ordered ones. All things being equal, therefore, the disordered states will prevail. Chaos will inevitably reign.

Physicists have a special way of quantifying disorder. They call it "entropy." The entropy of a broken coffee mug is a convenient measure of the total number of ways in which the mug can be broken. When a coffee mug breaks, and it becomes more disordered, physicists say that its entropy has increased. This statement has been generalized into one of the cornerstones of physics: the second law of thermodynamics. The law states that, overall, entropy can only increase or, at best, stay the same. It can never decrease.† The arrow of time, sometimes called the "thermodynamic arrow of time," is therefore associated with the direction in which entropy increases.

Initial/Final Conditions

All this reasoning about the direction of time assumes one thing. It assumes that a body—whether a coffee mug or a star or a human being—

---

* The term "arrow of time" was introduced by the English astronomer Arthur Eddington in his book *The Nature of the Physical World* (1928) (cited in Edward R. Harrison, *Cosmology* [Cambridge: Cambridge University Press, 1981], 144). "The great thing about time is that it goes on," wrote Eddington. "But this is an aspect of it which physicists seem inclined to neglect. . . . I shall use the phrase 'time's arrow' to express this one-way property of time which has no analogue in space."

† An equivalent statement of the second law is that heat cannot be turned into useful "work" without another change elsewhere. Work occurs when energy is concentrated in a few things: a single spring, a single piston. Heat represents energy dispersed among many, many things—for example, among the many molecules in a hot gas. The reason heat cannot be changed into work with 100 percent efficiency is that it is extremely unlikely that the energy in a gas will pass from the many, many states of a hot gas to the very few states of a moving piston. This does happen in a car engine. However, in that case there is a change elsewhere—the burning of the fuel, which is associated with an increase in entropy.

starts out in an ordered state but is absolutely free to end up in any final state possible. "This is the context of the second law of thermodynamics, which dictates that things decay and grow old," says Schulman. "But—and this is the crucial thing—there are other situations where the law has no jurisdiction."

Physicists call the kind of restriction that has a coffee mug intact at the outset an "initial condition." "It turns out that having an initial condition but no final condition positively guarantees that disorder will increase with time," says Schulman. "It is what creates the sequence of events we associate with the normal arrow of time."

But what if the constraint on a body is imposed at the end rather than at the beginning? What if there is a final rather than an initial condition? At first sight, this sounds completely ridiculous. In the case of the coffee mug, for instance, it is hard to imagine a constraint that would ensure that the broken coffee mug ended up intact.

However, future constraints are not really so bizarre as they seem. Think of a group of Londoners from all walks of life who sign up to learn Spanish at a night class once a week. During the week, they have different jobs and different circles of friends. But, though their lives lead them to different parts of the city, for the duration of the course there is a future constraint on their lives. When it comes to Friday at 7:00 P.M., they will all be in the same room learning Spanish.*

The Arrow of Time and Cosmic Expansion

Could our universe have a future condition imposed on it? The answer is: nobody knows. However, one way such a constraint might arise is if we lived in a Big Bang/Big Crunch universe. Such a universe, having expanded from a hot, highly compressed state in the distant past—a Big Bang—would one day shrink back down to a sort of mirror image of the Big Bang—a Big Crunch. If we lived in such a universe, the bodies inside the universe might be more limited in what they could do

---

* A physicist might object to this, saying that entropy relates to mindless entities rushing about randomly, not to "consciously directed" language students. So think of the forty students as forty atoms of a gas. During the week, the gas atoms fly about randomly, going to many places and giving the gas a high level of "disorder." But as 7:00 P.M. Friday approaches, a certain "order" begins to develop as first the velocities of the gas atoms orient themselves toward a single place and, finally, they all come together in an extremely "unlikely" cluster.

than we naively think. Galaxies such as our own Milky Way, for instance, may be constrained in the way they fly through space by a "future memory" of the universe squeezed back down into a tiny volume, much as the Spanish-language students are constrained by their need to attend their class on Friday night. "It is in situations where there is a final condition that disorder can actually change into order," says Schulman. "In other words, there can be a backward arrow of time!"

This possibility was first pointed out by the astronomer Thomas Gold of Cornell University in 1958. His argument turned out to be flawed, but the conclusion was confirmed by Schulman with more-rigorous reasoning in the 1970s. Schulman was able to show that, provided the Big Crunch at the end of the universe is a very ordered state, the arrow of time in a future contracting phase of the universe will point backward.* As the universe shrinks down to the Big Crunch, in other words, cool objects will grow hot, candles and stars will suck in light, and living things, having begun in the grave, will end up in the cradle.

Surprisingly, however, occupants of a future contracting phase of the universe will not see anything peculiar. "Because of their reverse arrow of time, they will see everything running backwards," says Schulman. "Consequently, they will actually perceive the contracting universe as expanding, exactly as we do."

The truly remarkable thing is that the arrow of time should be connected to what the large-scale universe is doing: growing or shrinking. Currently, the universe appears to be expanding, its constituent galaxies flying apart like pieces of cosmic shrapnel in the aftermath of the explosion of the Big Bang. "This is the ultimate reason that your coffee cools rather than warms up!" says Schulman.

Recall that the reason the arrow of time points the way it does for coffee mugs and the like is that they start out in an ordered state. It follows therefore that the reason the arrow of time points the way it does everywhere we have looked is that the universe itself must have started out in a highly ordered state. We can imagine that state. Because the universe is expanding, it must have been smaller in the past. If we imagine the expansion running backward, like a movie in reverse, we

---

* There is always the possibility that the Big Crunch may not be an ordered state. It may, for instance, contain large numbers of merging black holes, the relics of exploded stars. Black holes are associated with a lot of disorder. But nobody knows enough about the number of black holes there might be in a Big Crunch to answer the question one way or the other.

come to a time, roughly twelve to fourteen billion years ago, when everything in the universe was compressed into a tiny, tiny volume. This was the beginning of time—the Big Bang.

The ultimate reason why people grow old and buildings crumble is therefore that the universe at the time of the Big Bang was in an extraordinarily ordered state. Why was it in such an ordered state? If you can answer that one, there is a Nobel Prize waiting for you.

The Coexistence of Opposite-Directed Arrows

Surprisingly, none of this is particularly controversial. Physicists have long known that the arrow of time could point either forward or backward. It's simply a matter of initial or final conditions—whether there are past or future constraints. Until now, however, few considered the possibility of a backward arrow of time. The belief among physicists was that it was impossible for a region where time runs forward to coexist with a region where time runs backward. A region with backward-running time would simply be too fragile, too easily destroyed. "Think of a game of pool in which the balls are arranged in a triangular formation, struck by the cue ball and then scattered to the corners of the table," says Schulman. "Okay, now imagine the reverse-time scenario."

According to Schulman, if the balls are to follow the precise paths necessary to finish back in the triangle, a monumental amount of coordination will be needed. The slightest interaction with a region with normal time—for instance, the smallest cry of amazement from somebody watching the converging balls—could disturb the air enough to wreck everything. "For this reason people have argued that the backward arrow of a reverse-time region would be instantly destroyed by interacting with a normal-time region," says Schulman. "I've even heard people say that merely shaking an electron near the star Sirius would be enough to destroy a process with a backward arrow on Earth.*

---

* This is a variation on a statement first made by the French mathematician Emile Borel in 1924. In a book on probability, Borel calculated that moving one gram of matter by a few centimeters on Sirius would, when the effect of this reached Earth, completely change the microscopic state of a gas on Earth. This would happen, Borel pointed out, within the time it takes the atoms of such a gas to bounce off each other a few times, which is within a few billionths of a second.

"There is a fundamental flaw in this argument, however," says Schulman. "By assuming that the normal-time region can destroy the backward arrow, people are assuming that the normal-time region is in some way privileged. It isn't. The situation is perfectly symmetric. In reality, the reverse-time region is just as likely to destroy the arrow of the normal-time region as the other way around. All we can safely say is that, if the two regions interact, their arrows will either both be destroyed or both survive."

Most physicists would have put good money on the first possibility. The idea that regions with opposite arrows of time could coexist was simply too absurd to contemplate. However, Schulman, with the aid of a simple computer model, has shown that most physicists are wrong. As long as the two regions interact only weakly, both arrows of time can survive.

Schulman's computer model, he maintains, captures the essential features of a gas of particles flying about in a closed box—or rather, two distinct gases, each in its own box. In the first, the gas particles are constrained to start off in one corner. As time passes, they gradually spread out to fill the whole box. This change from order to disorder is the hallmark of a normal arrow of time. In the second box, however, the particles start off all over and are constrained to end up in a corner. Because the gas goes from a disordered state to an ordered one, it possesses a reverse arrow of time.

Having set up two gases with opposite arrows of time, Schulman simply arranges for them to interact. In other words, he sets things up in his computer model so that what happens to one gas has a slight effect on what happens to the other.

According to the received wisdom, any interaction between the gases should wipe out both arrows of time. That is, both gases should end up in a spread-out state so disordered that there is no further change in the state with time. A gas in such a state actually has no arrow of time. What Schulman finds instead is that neither arrow is destroyed. Contrary to all expectations, both arrows are robust enough to survive a weak interaction. The implications are remarkable. "There could be places within our universe—perhaps not too far away—where time runs backwards," says Schulman. "Places, in other words, where eggs unbreak, milk stirs itself out of coffee and people grow younger with every passing year!"

Seeing a Region with Backward-Running Time

Schulman's discovery that regions with opposite arrows of time can survive a mild interaction without losing their arrows is remarkable
enough. However, he has also demonstrated something else extraordinary. "People in such regions could actually see each other!" he says.

To ask whether we could see a reverse-time region, and vice versa,
amounts to asking whether it is possible to describe light bouncing
back and forth between the regions in a mathematically consistent—
that is, noncontradictory—manner. Astonishingly, Schulman finds that
there is such a way. "It is perfectly possible for us to see a reverse-time
region, where light goes back to its source, and for people in a reverse-
time region to see us," he says. "For each of us, there is perfectly normal transmission and reception of light. We would see them at
successively earlier times by our clocks and they would see us at successively earlier times by their clocks."

In Martin Amis's powerful and disturbing novel *Time's Arrow,* a Nazi
war criminal regresses through his life to the scene of his terrible
crimes. For Amis, the reverse-running river of time was merely a liter-
ary device, which endowed the harrowing story with a horrible momentum as it raced back to its inevitable conclusion at Auschwitz.
Schulman's discovery is that, if a world like Amis's really did exist, there
would be nothing in principle to prevent us from observing it from a
distance. "We could see history playing backwards exactly like a movie
in reverse," he says.

But although Schulman has shown that nothing untoward would
happen if you looked at a reverse-time region, there are still some sticky
paradoxes with which to grapple. For instance, it is possible to imagine
the following situation. A normal-time observer—let's call her Alice—
can see through an open window that rain has wet the carpet of someone in a reverse-time region—call him Bob. Alice could wait until
before the rain started and shout to Bob to close his window. But, if he
does, and his floor doesn't get wet, Alice wouldn't be moved to shout
her warning. . . . So, did Bob's floor get wet or not? Schulman's guess is
that the "carpet paradox" goes away if the problem is posed carefully.

One possibility, he says, is that the paradox is created entirely by the
initial and final conditions. "If you impose such conditions, it may
turn out that the events described simply cannot happen—in mathematical terms, the events are not a solution of the problem," he says.

"Clearly, if the events are forbidden from happening, there is no paradox to worry about."

Another possibility, according to Schulman, is that something intermediate happens to "smear out" the paradox. "Perhaps Alice sees the window open, shouts to Bob but the message gets blurred and Bob doesn't close the window completely," says Schulman.

One scenario is the following. "Alice looks and sees Bob's window slightly open with a small trickle of water on the carpet," says Schulman. "Should she contact him or not? She dithers a moment, then decides to send a message. However, it is rather unassertive: 'If it's going to rain, you might want to close the window.' Bob, who likes fresh air, decides that the message is not all that compelling. He therefore decides not to close the window fully, but to leave it open a crack, despite the predicted rain. This is the self-consistent solution."

How Could a Backward Arrow Have Arisen?

The discovery that opposite arrows of time can coexist within our universe opens up the possibility that there exist regions relatively nearby where time runs backward. It is even possible, says Schulman, that such regions exist within our own galaxy, the Milky Way. The question then arises: how could they have come about? After all, even if we do live in a Big Bang/Big Crunch universe, reverse-time regions will surely be found only in the contraction phase of the universe, and that exists only in the far, far future.

However, even here our naive expectations may be incorrect. According to Schulman, in a Big Bang/Big Crunch universe, it would be perfectly possible for relics from our forward-time, expanding phase of the universe to survive into the contracting phase. "Equally, relics from the reverse-time, contracting phase could survive into our expanding phase," says Schulman.

This highlights an important point about the arrow of time in a Big Bang/Big Crunch universe. Just because time flows forward in our expanding phase and backward in a future, contracting phase, it does not mean that, at the "turnaround" point when the expansion of the universe finally runs out of steam and contraction takes over, the direction of time suddenly flips. If this happened, any exploding stars would immediately start unexploding, any living things would stop aging and immediately begin regressing to their births. No, nothing like this happens. According to Schulman, the turnaround is more patchy, with the opposite arrow of time establishing itself only gradually. "Instead of flipping suddenly, the arrow would change over a very long time," he says. "Conscious beings would not notice any change unless they had memories roughly as long as the lifetime of the universe."

The very gradual nature of the change from one arrow to another is the reason that isolated regions could maintain their forward arrows far into the contraction phase. Similarly, and this is the most mindblowing thing, isolated regions could maintain a backward arrow far into our expansion phase. If there were sentient beings and they kept records that fell into our hands, we could "know" our own future.

If the idea of relics from the future surviving into the past sounds nonsensical, remember that, from the point of view of anyone in the contracting phase and experiencing a reverse arrow of time, such regions would be surviving into the future. "There is perfect symmetry here," says Schulman.

If we do live in a Big Bang/Big Crunch universe, the turnaround point could be more than a hundred billion years away. This is much greater than the ten-billion-year-or-so lifetime of a star such as the Sun. Consequently, any reverse-time relics from the far future that have survived until today will come from a time when all the bright stars have long ago run out of fuel and burned out. The likelihood is therefore that such a relic region will be dark. Although we would love to see amazing things such as stars unexploding, we may only be able to detect such a region by the gravitational pull it exerts on visible stars and galaxies. Not very exciting, you might think. However, this raises a remarkable possibility.

Roughly 90 percent of the matter in the universe is known to give out no light whatsoever. We know of the existence of this "dark matter" only because of its gravity, which pulls on the visible stars and galaxies and changes the way they fly through space. There have been hundreds of proposals for the identity of the universe's mysterious dark matter. But, almost certainly, none is so bizarre as that of Schulman. "Some of the universe's mysterious dark matter could be reverse-time matter from the future," he says. "I must stress that this is a very, very speculative possibility."

Another, only marginally less off-the-wall possibility is that in the twelve to fourteen billion years since the Big Bang, most of the matter in the universe—90 percent, to be exact—has actually collided with reverse-time matter from the future. "Such collisions would create matter in 'equilibrium'—matter, that is, with no time direction at all," says Schulman. "Once again, it would appear exactly like dark matter."

A Temporal Cosmological Principle

But say we do not live in a Big Bang/Big Crunch universe. The latest astronomical observations appear to indicate that the combined gravity of all the galaxies pulling on each other would simply not be enough to brake and eventually reverse the universe's expansion. If this is correct—and nobody can say for sure—then the galaxies will go on flying apart from each other forever. There will be no Big Crunch at the end of time. "However, a universe with a future contraction phase is only one possible way in which a reverse arrow of time might arise," says Schulman. "All that is really needed is a final condition—a future constraint on the state of the universe."

This could arise, according to Schulman, in a way we cannot at present imagine. After all, we have only a hazy idea of the ultimate laws of physics and so cannot predict with any certainty the future of the universe.* "Perhaps regions with backward arrows arise for the same mysterious reason that forward arrows arise," says Schulman. "After all, we do not yet know why the universe started out in such a highly ordered state, the necessary condition for a normal arrow of time."

This prompts the observation that maybe there is nothing special at all about our direction of time. We have come to think of the direction as normal simply because it is the way it is in the region of the universe where we happen to live. "Maybe at some fundamental level the universe permits both directions of time," says Schulman.

Certainly, Schulman has shown that the two arrows of time are in many respects equivalent. After all, people in a contracting phase of the universe would experience a reverse arrow of time and therefore see the contraction as expansion, just as we do. What from our perspective

---

* If there is a future constraint on our universe, an exciting possibility is that it might actually be discernible today. For instance, if we lived in a universe that would ultimately recollapse to a Big Crunch, the galaxies would move in a way subtly different from the way they would move if they had complete freedom in the future.

is the Big Crunch at the end of time would from their perspective be the Big Bang at the beginning of time. Conversely, our Big Bang would be their Big Crunch. We might then ask how we can be sure we are not in a contracting phase. How do we know that our coffee is cooling down rather than warming up? The simple answer is: we can't know. If Schulman's work tells us anything, it tells us that there is nothing special about a forward arrow of time. Things would very likely appear the same with a backward arrow.

It seems the more we learn about the arrow of time, the more we realize that we are only just beginning to come to grips with the whole concept. That sentiment is echoed by the mathematician Gregory Chaitin of IBM's Thomas J. Watson Research Center in Yorktown Heights, New York: "Despite many statements to the contrary by distinguished physicists, I've always been convinced that nobody quite understands the arrow of time."

Schulman's suspicion remains that there is no preferred arrow of time. If he is right, the future is not what it used to be. In *The Go-Between*, L. P. Hartley wrote: "The past is a different country. They do things differently there." But maybe Hartley was wrong. History assumes a totally different meaning if it can be either past history or future history.

Such a redefinition of history is enough to leave your head spinning. However, it turns out to be nothing compared with the redefinition suggested by a young Swedish physicist named Max Tegmark. For if Tegmark is right, there is not just one history but an infinity of different histories, each peopled by a different version of you.

2

I'm Gonna Live Forever

Evidence is growing that there are an infinite number of realities stacked together like the pages of a never-ending book

The quantum theory of parallel universes is not some troublesome, optional interpretation emerging from arcane theoretical considerations. It is the explanation—the only one that is tenable—of a remarkable and counter-intuitive reality.

—David Deutsch, *The Fabric of Reality*

Whenever a creature was faced with several possible courses of action, it took them all, thereby creating many . . . distinct histories of the cosmos. Since in every evolutionary sequence of the cosmos there were many creatures and each was constantly faced with many possible courses, and the combinations of all their courses were innumerable, an infinity of distinct universes exfoliated from every moment of every temporal existence.

—Olaf Stapledon, *Starmaker*

Picture this scene:

A spotless laboratory, empty except for a young woman, dressed in a white coat, an elderly man, and a machine gun set up in the center of the floor. The woman stands at a control panel, her index finger trembling above a red button. The man, who is trying desperately to look confident, lifts his hand periodically to wipe away the sweat from his forehead.

The machine gun is set up so that each time it comes to fire it can either fire or not fire, in which case it merely makes a dull click. Whether it fires or clicks is determined completely randomly. Think of a coin being tossed: heads it fires, tails it clicks.

15

"Okay, I'm ready," says the man, stepping into the machine gun's firing line. The woman hesitates. "Go on," he urges. "We agreed before. I'm old. I've nothing to lose."

Biting her lip, hardly daring to look at the man, she forces herself to press the red button.

Cut to woman's point of view: Click.

Cut to man's point of view: Click.

Cut to woman's point of view: Click.

Cut to man's point of view: Click.

Cut to woman's point of view: Bang! She screams and rushes to the man, who now lies dead in a pool of blood.

Cut to man's point of view: Click. Click. Click again. Ten more clicks. Thirty. Finally, on the hundredth click, the man steps out of the firing line. Smiling triumphantly, he embraces his relieved assistant. "See, the theory is right," he says. "I am immortal!"

Confused? You should be. After all, from the woman's point of view, the man was killed by a bullet after two clicks of the machine gun, whereas from the man's point of view he heard a hundred impotent clicks and stepped out of the firing line totally unscathed. How could the man have been both killed and not killed? There is only one way: if there exists more than one reality.

Suppose that, each time the machine gun comes to fire a bullet, it both does and does not. In other words, the universe splits into two entirely separate realities—one in which the woman sees the man shot dead, and another in which an identical version of the woman sees an identical version of the man survive. The next time the gun comes to fire, it both does and does not again, and the universe splits into two more realities, with two more versions of the man and the woman. And so on.

It was to test his belief that multiple realities exist that the elderly man set up the death-defying experiment. If he was wrong and there exists only a single reality, by stepping in front of the machine gun he would be committing certain suicide. Hence the beads of perspiration. Hence his advanced age. This isn't the sort of experiment you want to risk until most of your life is behind you; until, as the man said, you have "nothing to lose." If, however, the man was right and multiple realities do exist, there would always be realities in which a version of himself heard nothing but dull clicks, no matter how many times the gun came to fire.

"Of course, in most realities the man would indeed die and his woman assistant would scream in horror," says Max Tegmark, the physicist at the University of Pennsylvania who proposed this bizarre experiment. "But—and this is the key point—he would have no aware-ness of these realities since he would be dead! The only realities he would continue to be aware of would be the ones in which he survived. It was inevitable therefore that after fifty clicks, a hundred, two hun-dred, he would step out of the firing line of the machine gun having discovered he was immortal!"

But surely the idea of multiple realities in which all possible alter-nate histories are played out is pure science fiction? "Not at all," says Tegmark. "An increasing number of physicists are coming to believe it." The reason? It may explain one of the greatest mysteries of modern-day science: why the world of atoms behaves so differently from the everyday world of people, trees, and tables.

More Places than One

The theory that describes basic building blocks of matter—atoms and their constituents—is called quantum theory. It is fantastically success-ful; it is arguably the most successful scientific theory ever devised. Its predictions have been verified to many, many decimal places in thou-sands of experiments. Quantum theory has made the modern world possible. It has led not only to computers, lasers, and nuclear reactors but also to an explanation of why the ground beneath our feet is solid and why the sun shines.

The successes of quantum theory are so numerous and so spectacu-lar that there can be little doubt that it encapsulates a great deal of truth about our universe. However, quantum theory tells us something rather disturbing about atoms and their like: they can be in several places at once.

This is more than mere theory; it is actually possible to observe the consequences of an atom being in two places at once. In a well-known physics experiment, for instance, particles from the atomic realm, such as photons of light or electrons or fully fledged atoms, are fired like bul-lets at a wall with two closely spaced vertical slits cut into it. All the evi-dence from countless runs of this "double-slit" experiment indicates that each individual particle passes through both slits simultaneously.

This bizarre property of atoms poses a big puzzle for physicists. How is it that atoms can be in many places at once, whereas tables, trees, and pencils—big things made out of atoms—quite emphatically cannot? Reconciling the difference between the "microscopic" world of atoms and the "macroscopic" everyday world turns out to be the central problem in interpreting quantum theory.

The standard explanation, which has been widely accepted ever since the birth of quantum theory in the 1920s, merely concedes that the world of atoms and the world of tables are radically different. It then concocts a recipe for translating from one domain to the other. According to the "Copenhagen interpretation," the act of "observing" a microscopic particle such as an atom is crucial. It forces it to make up its mind about where it wants to be, and opt for being in one place and one place only out of all the possible places it could be. Lunatic behavior is, therefore, tolerated until the instant of observation. Thereafter, a particle behaves as a well-defined, localized object.*

In the case of the double-slit experiment, the Copenhagen interpretation tells us that the particles can go through both slits in the wall simultaneously only as long as they are not observed. The instant an experimenter puts some kind of detector next to the slits to try to discover which slit a particle goes through, this act of observation will force the particle to make up its mind and choose one slit over the other.

The Many Worlds

In effect, the Copenhagen interpretation says that quantum theory works for describing small things such as atoms but not for describing big things such as tables. However, there is another, arguably simpler way of reconciling the difference between the microscopic world of

---

* The Copenhagen interpretation is itself open to interpretation because of its vagueness about what constitutes an "observation." For some people, an observation involves a macroscopic object such as a particle detector interacting with a microscopic object such as an atom. For others, it involves an interaction with some kind of conscious being. According to this extreme view, atoms do not exist as well-defined, localized objects until someone looks at them. Stars and galaxies become real only when their light is picked up by our telescopes. The universe has been waiting for us to come on the scene and, by observing it, bring its past and present into being.

atoms and the macroscopic everyday world. It was proposed by a Princeton graduate student named Hugh Everett III in 1957. "According to the 'Many Worlds interpretation,' set out by Everett in his Ph.D. thesis, quantum theory applies not simply to atoms but to everything," says Tegmark. "The world of tables, in other words, is exactly the same as the world of atoms."

Surely Everett was not seriously saying that tables can be in many places at once? "Yes, this is exactly what he was saying!" says Tegmark. But that's crazy; nobody has ever seen such a "schizophrenic" table. "Yes, that's perfectly true," says Tegmark. "And Everett had an explanation for that as well. If you observe a table which is in two places at once, your mind will also end up in two states at once—one state which perceives the table in one place and one which perceives it in another place!"

In effect, there will be two versions of you, each perceiving a world in which there is a different version of the table. "This is why people call Everett's idea the Many Worlds," says Tegmark. "A more accurate name, however, might be the Many Minds."

Think of a quantum system such as an atom as represented by a bundle of threads. In the Many Worlds model, separate threads within the bundle correspond to separate perceived realities. In the Copenhagen picture, however, the bundle is a rather ghostlike amalgam of possibilities until an act of observation singles out one thread to be made solid and real.

Abandoning the idea that there is a single reality and accepting multiple realities instead is a big step to take. If Everett was willing to take it, there must surely have been a big payoff. It turns out that there is. The Many Worlds idea makes sense of many utterly baffling features of the world. Take the case of microscopic particles going through slits in a wall. Here, the evidence that each particle goes through both slits comes from a distinctive pattern the particles make on a flat screen placed on the far side of the wall. It is known as an "interference pattern."

Some regions of the screen are hit often by particles, and other regions are never hit. Imagine that the places that are hit show up black and that the places that are not hit show up white. After the experiment has been run for a while, and a large number of particles has passed through the slits, an experimenter will find that the screen is covered in vertical stripes, which alternate black and white. The alternating stripes are the interference pattern.

To make such a pattern, particles that pass through one slit have to mingle, or "interfere," with particles that pass through the other slit. The pattern simply does not form if the particles go through one slit only. Well, the amazing thing is that an interference pattern forms even if particles are shot at the wall one at a time, with long gaps of time between them. Must we accept that a single particle can somehow combine with itself?

The answer, according to the Many Worlds idea, is no. A single particle does not interfere with itself. That is nonsensical. No, what it does is interfere with a particle going through the other slit. Wait a minute, what other particle? The answer is as simple as it is mind-blowing. Another particle in a neighboring reality. According to the Many Worlds hypothesis, this simple tabletop experiment, which even students can set up, is providing us with earth-shattering evidence of the existence of other realities.

The double-slit experiment highlights a crucially important property of the multiple realities of the Many Worlds idea. When reality splits, the new branches do not go their separate ways, having nothing to do with each other ever again. At least, they do not have to. Instead, after splitting, parallel realities can later come back together and interfere with each other. In the case of the double-slit experiment, for instance, the particles in each branch of reality—the one in which the particle goes through one slit and the one in which it goes through the other slit—come back to interfere with each other and produce the pattern on the screen. "If it was not for this interaction between branches of reality, the Many Worlds would be very dull—and explain nothing," says Tegmark.

But although the different branches of reality interact with each other, they do so in a limited way. For instance, the interaction of realities has no obvious consequences for the everyday world. Nor are there any convenient home experiments that demonstrate the phenomenon. But then, it is a general feature of quantum theory that its bizarre effects have essentially no manifestations in the large-scale world.

The explanatory power of the Many Worlds idea really comes into its own in the case of "quantum computers." In these devices, which are currently creating much excitement in the world of physics, experimenters cleverly exploit the ability of particles such as atoms to be in many places at once to do many calculations at once.

As yet, the field is in its infancy, and the most powerful machines built to date are able to process only a handful of binary digits, or "bits," compared with the hundreds of millions of bits routinely processed by conventional computers. However, there is no reason to believe that far more powerful quantum computers will not be built within the next ten or twenty years. Such computers, in a sense the ultimate computers, would make conventional computers appear positively retarded. They could solve in seconds certain problems that would take a conventional computer longer than the age of the universe. But this barely hints at the awesome potential of quantum computers. "If you imagine the difference between an abacus and the world's fastest supercomputer," says science writer Julian Brown, "you would still not have the barest inkling of how much more powerful a quantum computer could be compared with the computers we have today."

And therein lies one of the central problems of quantum computers: explaining their phenomenal capabilities. If we look to the future, it is easy to imagine better and better quantum computers, which are capable of doing more and more simultaneous calculations. Extrapolating into the far future, there is nothing in principle to rule out the possibility of a quantum computer so powerful that it could carry out more calculations at any one time than there are fundamental particles in the universe. Fundamental particles are the ultimate building blocks of the universe, the constituents of atoms. A rather interesting question now arises. As the Oxford physicist David Deutsch has asked: where would such calculations be carried out? After all, if a quantum computer were doing more calculations at any instant than there are particles in the universe, the universe simply would not have the physical resources at its disposal to do what the computer was doing.

The Many Worlds idea, according to Tegmark and others, provides a natural, but nonetheless staggering, answer to the conundrum. A quantum computer is never short of the resources it needs to carry out its calculations because it does not have to rely on a single universe. "Different parts of its calculations are in a sense performed in different realities," says Tegmark. "Bizarre as it seems, quantum computers achieve what they achieve by exploiting huge numbers of versions of themselves in other neighboring realities."

Imagine, for a moment, that you have a quantum laptop and you assign it a problem. In effect, infinite versions of your laptop work on parallel strands of the problem. Finally, a split second later, those strands are pulled together to provide the answer that pops up on your screen.

There is a big caveat. A normal computer, programmed to carry out simultaneous, or "parallel," calculations, is free to produce the result of all the calculations at the end. A quantum computer, however, can produce only a single result in a special form. "Because of this, people have to work extremely hard to come up with recipes for calculating useful things," says Tegmark. "It's important to bear this in mind and not exaggerate the capabilities of quantum computers."

There is no doubt, however, that quantum computers could manipulate information at an unprecedented rate. However, this is not what is most amazing about them. "What is most amazing about them is that they are the first machines humans have ever built that harness the cooperation between different universes," says Tegmark.

Why Do We Experience Only One Reality?

The Many Worlds idea helps us explain the behavior of quantum computers and microscopic particles flying through slits in a wall. But, paradoxically, these systems also highlight a serious problem with the idea. The Many Worlds explanation hinges on the ability of quantum computers and microscopic particles directly to experience infinite parallel realities. But if quantum computers and atoms do indeed live in this world of multiple realities, why do we human beings only ever experience one reality?

Because there was no obvious answer to this question, the Many Worlds idea was largely ignored for several decades. It wasn't simply that physicists recoiled in horror from the nightmarish proliferation of realities—although this definitely played a part. Part of the reason for the physicists' reluctance to embrace the Many Worlds idea was the sticky problem of explaining why none of us—except perhaps those suffering the effects of alcohol or concussion—actually perceives multiple realities. The Many Worlds hypothesis, after all, explicitly says that atoms and pencils and tables can be in many places at once. It goes so far as to predict that, if a pencil is balanced exactly on its tip, it should fall down in all possible directions simultaneously.

In 1957, when he proposed the Many Worlds idea, Everett was completely unable to explain why we do not see pencils falling in all directions at once. The explanation had to wait until the 1970s and 1980s, when Dieter Zeh of the University of Heidelberg in West Germany and Wojciech Zurek at the Los Alamos National Laboratory in New Mexico began to investigate a phenomenon with the impressive-sounding name "environmentally induced decoherence," or simply "decoherence."

The crucial idea is that "superpositions," in which atoms and their like can be in several states at once, are extremely fragile. They can exist on one condition and one condition only—that the object in the multiple state is totally isolated from its surroundings. The merest interaction with the outside world will destroy the superposition through decoherence.

According to Zeh and Zurek, the interaction with the outside world need be only tiny. Even a single photon of light bouncing off the pencil standing on its tip turns out to be enough to shatter its isolation, causing it to fall in just one direction. The photon takes information about the pencil to the rest of the world. "It's almost as if schizophrenic quantum behavior is a secret," says Tegmark. "But once the outside world learns the secret, the secret is no longer a secret. Consequently, the weird behavior no longer exists."

A big thing, such as a pencil, is constantly jostled by air molecules and photons of light. Consequently, the quantum secret leaks into the outside world fantastically fast. If a pencil balanced on its tip happened to be in a superposition in which it started falling in all directions at once, it would decohere in an interval of time so short it would make the blink of an eye seem like an eon. Before any mental process could register the pencil in its multiple state, the pencil would be transformed into a perfectly normal pencil falling in just one direction.

According to Zeh and Zurek, whether an object can appear to be in many places at once depends entirely on whether it is isolated from its surroundings. It is not true that weird quantum behavior is intrinsically a property of small things such as atoms but not big things such as pencils and tables. The reason that small things can exhibit quantum behavior, appearing in several places at once, whereas big things generally cannot, is that in practice it is easier to isolate a small thing than a big thing from its surroundings. This truth is the bane of the life of those struggling to build quantum computers. Whether they succeed or fail in exploiting the multiple nature of atoms depends entirely on how well they can isolate their machines from their surroundings.*

The key idea of decoherence, then, is that the collision of an object with a single photon or single molecule can destroy its quantum nature. It seems unbelievable that such a small event can have such a catastrophic effect (or, alternatively, that quantum objects are so incredibly fragile). So what exactly does a photon or molecule actually do to an object to blow its quantum nature to smithereens? It's a key question. But here, the physicists are of no help. "I'm sorry but I simply don't have a nonmathematical explanation," admits Tegmark. "If you find one, please let me know!"

Decoherence is a practical matter. It prevents us from perceiving big things, such as pencils, in many places at once because we have trouble isolating them from their surroundings. However, it does not prevent us in principle—and maybe one day not even in practice. After all, we are getting better at isolating things from their surroundings. Anton Zeilinger's team at the University of Vienna is already talking about sending viruses through a double slit, raising the possibility of a relatively big object being in two places at once. What if that team succeeds with something big enough to see with the naked eye? Would we then see it in a superposition of all its states? Would Zeh and Zurek's fix turn out to be merely a temporary fix?

The answer, according to Tegmark, is no. It is commonly believed that brain cells known as neurons play a central role in conscious thought by transmitting electrical pulses, a process known as "firing." For an object to be perceived as being in two places at once, according to Tegmark, at least one neuron in the brain would have to be in a state in which it was simultaneously firing and not firing. However, he points out that such a superposition could not stay isolated for long in the messy environment of the brain. Within a fantastically small time, a water molecule or something else would bump into the neuron and destroy the superposition. "The interval is so tiny that no mental

* Skeptics say it will be impossible to build fully fledged quantum computers because it will be impossible to isolate them completely from their surroundings. Recently, however, physicists have thought up ingenious ways of correcting the errors introduced as the components of a quantum computer inevitably decohere, losing some of their quantum nature. Remarkably, it appears that this error correction can always stay one step ahead of the game, so that a quantum computer always gets to complete its calculation before it stops being a quantum computer.

process could register it," says Tegmark. "Hence we wouldn't see a big object in several places at once even if it were possible to isolate it from its surroundings."

This leaves only the question of normal thought processes creating superpositions of mental states. Might it be possible, for instance, to be both hungry and not hungry at the same time? "Once again, decoherence in the brain is far faster than any conceivable mental process so the brain is protected from weird superpositions," says Tegmark.

Dying to Know: Testing the Many Worlds Idea

Decoherence, and its explanation of why we do not perceive multiple realities directly, is one of the principal reasons a lot of physicists are now willing to "come out of the closet" and embrace the Many Worlds idea. In July 1999, at a conference on quantum computation held at the Isaac Newton Institute in Cambridge, England, Tegmark carried out an informal poll to find out what interpretation of the quantum theory was favored by the participants. The result was surprising. "For the first time in more than seventy years hardly anyone supported the standard Copenhagen interpretation," says Tegmark. "But the stunning thing was that the main challenger was the Many Worlds. In fact, it was almost ten times as popular."

In other words, physicists are increasingly accepting the idea that there are infinite realities stacked together like the pages of a never-ending book. So there are infinite versions of you, living out infinite different lives in infinite parallel realities. In some of the realities, you never opened this book; you never started to read these words. In other realities, you had an entirely different upbringing, developed radically different interests, made completely different friends. "In the Ultra-verse," says novelist Martin Amis, "there is an infinite number of universes and an infinite number of planets, and in infinity everything recurs an infinite number of times!"*

Neighboring realities are similar to one another. For instance, two could differ by as little as the fact that in one an experiment is performed in which an atomic particle passes through the left-hand slit of a wall and in the other it passes through the right-hand slit. For billions

---

*   Martin Amis, *The Information* (London: Harper Collins, 1996), 436.

of years before this trivial event, the two realities could have shared identical histories.

Realities that are far apart can be quite different. Somewhere in the infinity of realities, for instance, there are Earths that were not devastated by a comet's impact sixty-five million years ago and on which the dinosaurs evolved into intelligent beings. There are Earths where the Industrial Revolution started in China, not in Britain, where Marilyn Monroe married Einstein, and where the Nazis prevailed in the Second World War. Just such a future—in which the Germans occupy the Atlantic side of the United States and the Japanese occupy the Pacific side—was depicted in Philip K. Dick's classic science fiction novel *The Man in the High Castle*.

But the Many Worlds idea, its proponents claim, is far from science fiction. It is a vision of ultimate reality, and it could be tested. At least, this is Tegmark's contention.

In his proposed experiment, the firing of a capricious machine gun is controlled by the "quantum spin" of a microscopic particle such as an electron. The quantum spin of an electron can point in one of only two possible directions, conventionally referred to as "up" and "down." The spin has a 50 percent chance of being up and a 50 percent chance of being down. It's a bit like the flip of a coin, except for a subtle difference. Whether the coin comes down heads or tails is in principle predictable if you know the direction and force of the coin flick, the coin's trajectory through the air, and so on. By contrast, quantum spin is inherently unpredictable—a truly random coin toss.

Tegmark suggests setting up the machine gun in such a way that if the spin of the particle is determined to be up, it fires a bullet; if the spin is determined to be down, however, the gun simply makes a click but does not fire.

The researcher has an assistant. She knows that her boss has a 50 percent chance of surviving after one spin measurement, a 25 percent chance of surviving after two, a 12.5 percent chance after three, and so on. "After ten measurements, her boss's chances of survival are less than one in one thousand," says Tegmark. "Or, to put it another way, he has a 99.9 percent chance of being dead."

This is the conventional view. However, if the Many Worlds idea is correct and all possible outcomes actually happen, says Tegmark, things look rather different to the researcher. A dead person perceives nothing with 100 percent certainty, whereas a live person perceives the world with 100 percent certainty. So no matter how many measurements are made, there will always be realities among the infinity of realities in which the researcher continues to perceive. "If the Many Worlds idea is correct, the researcher will hear a constant series of impotent clicks," says Tegmark. "He can stay in front of the machine gun as long as he likes. He will feel immortal!"

However, there is a catch. The researcher, having proved to his own satisfaction that the Many Worlds hypothesis is correct, will never be able to convince anyone else of this. Take the case where he waits for ten spin measurements. In 99.9 percent of the parallel realities, his assistant sees him die. And even in the one reality in which she sees him survive, she will argue that her boss was merely lucky, knowing that it will be necessary only to make him stand in front of the machine gun for a few more spin measurements to see him shot dead.

If the Many Worlds idea is correct, proving that it is correct is not particularly hard—if you don't mind your friends and loved ones in most realities seeing you commit suicide. This is an important observation. It turns out that, as far as predicting the outcome of any known experiment is concerned, the Many Worlds hypothesis and the Copenhagen interpretation are indistinguishable. They are among about half a dozen alternative interpretations of quantum theory, all of which predict precisely the same thing.

The reason all these equivalent interpretations exist is that quantum theory is written in an abstract mathematical language. For it to make contact with the everyday world, it must be translated into human language, which is more relevant to everyday experience. In everyday language, we often have several alternative ways of saying the same thing—for instance, we can say a glass is half full or half empty. Similarly, there are several different ways of interpreting the underlying mathematical reality of quantum theory. The fundamental problem physicists have to wrestle with is why we perceive objects as being in one place at a time rather than in many. The different interpretations of quantum theory are merely rival attempts at providing an answer. But if they are equivalent, why prefer the Many Worlds hypothesis to any of the others?

There are several answers. For a start, there is Occam's razor, a piece of ancient wisdom that says that, if there are several possibilities, one should always choose the simplest. And the Many Worlds idea is arguably simpler than the alternatives. After all, it states that quantum theory describes everything, not just the microscopic realm of atoms. Another reason for preferring the Many Worlds hypothesis is that it provides a better explanation—some would say the only reasonable explanation—of how atoms can be in two places at once and how quantum computers can do the astonishing things they can do. And if these reasons don't cut any ice, there is always Tegmark's experiment.

Will anyone ever do it? It will take a brave person. Tegmark's faith in the Many Worlds idea is deep. However, it turns out, it is not so deep that he would go to that extent. "I'd be okay but my wife, Angelica, would become a widow," he says. "Perhaps I'll do the experiment one day—when I'm old and crazy."

Human Consequences of the Many Worlds Idea

Tegmark's proposed experiment with the "quantum machine gun" illustrates one of the most remarkable consequences of the Many Worlds hypothesis: if you die in one reality, you live on in others. And because the only realities that you can possibly perceive are the realities in which you continue to live (naturally enough), you will discover as you live your life that you are miraculously immune to all life-threatening accidents. If a building collapses, you will find yourself crawling from the rubble; if a plane crashes or a bomb explodes, you will be the only survivor. Cats will have nothing on you. The word "lucky" will have to be redefined even to remotely approximate the extraordinary level of your good fortune.

But your amazing luck will be noticed only in an infinitesimal subset of realities. In the huge majority of realities, people will see you succumb to one accident or another. In this way the Many Worlds idea remains perfectly compatible with our everyday experience of a world in which good fortune occurs with low probability and few people live miraculously lucky lives.

If the Many Worlds hypothesis is correct, you will not die at age 60 or 70 or even 80. There will be a few realities in which you will live to the theoretical maximum possible age for your biological makeup. You will end up one day in a reality in which you are 110 or 120; it is inevitable. Such a reality is waiting for everyone, one for each of us. But why will you live until only 110 or 120? There must be realities out there in which someone discovers a cure for aging or in which someone discovers how to "download" consciousness into computers that live forever. Because these are the only realities you will continue to perceive, these will be the realities in which you will continue to live. Perhaps the fate of each and every one of us is to see the curtain come down on creation, to be spectators at the end of time.

If the Many Worlds idea is correct, it seems the only certainty in life is not death but immortality. We're all gonna live forever.

But let's not get too carried away. Most accidents that befall people are not of the simple "quantum" type envisaged by Tegmark in his experiment. The events that result in death are gradual and not the result of a single flip of a quantum coin. If the Many Worlds idea cannot legitimately be used to describe them, then the prediction that we all will survive until the end of time may be a prediction too far. "Whether the Many Worlds implies subjective immortality has bothered me for a long time," admits Tegmark. "I agree that if the argument is correct, I should expect to be the oldest guy on the planet, which I am not, severely discrediting Everett's idea."

The flaw in the argument, says Tegmark, is that dying is not a black-and-white issue, in which you are either dead or alive. Instead, a person goes through a whole continuum of states of progressively decreasing self-awareness. "I suspect that when I get old, my brain cells will gradually give out—indeed, that's already started happening—so that I keep feeling self-aware, but less and less so," says Tegmark. "My final 'death' will then be quite anticlimactic, sort of like when an amoeba croaks."

Successful quantum suicides, according to Tegmark, need to satisfy three criteria. First, the process that leads to death must be a truly quantum process so that the person involved truly enters a superposition in which he or she is both dead and alive. Second, the person must be killed quickly enough that he or she does not become aware of the outcome of the quantum coin tosses. Otherwise, for a second or two, there would be a very unhappy version of the person, who knows for certain that he or she is about to die. This would spoil the whole effect. Lastly, the person must be risking real death, not just injury.

So, what about most real-life accidents or diseases? Tegmark maintains that these do not satisfy either criterion 2 or criterion 3. What about cancer? A normal cell can be turned into a cancer cell, multiplying out of control, by a genetic mutation. And this could involve a single microscopic particle such as a photon or an electron interacting

with an atom in a DNA chain.* On this single microscopic event
hinges the fate of a human being—whether he or she will develop
cancer and die, or not develop cancer and live. It might seem that, in
this case, the Many Worlds hypothesis says unequivocally that the person concerned will continue to live in a reality in which cancer is
never developed. However, Tegmark believes this satisfies criterion 1
but not criterion 2.

Tegmark does not rule out immortality. "However, it may require
quite contrived circumstances such as those in the quantum suicide experiment," he says.

Death or immortality apart, what are the implications of the Many
Worlds idea for all the other versions of you? If you are fortunate
enough to have had a happy life, then it is likely that in the overwhelming majority of realities the other versions of you are living out
miserable lives. Perhaps they never met the person you fell in love with
and married. Perhaps they were never born into the loving family into
which you were born. Think of all the strokes of luck you have had in
your life and think of all the other yous that were not so lucky.

This idea was explored by the science fiction writer Larry Niven in
his story "All the Myriad Ways."† In Niven's world, the Crosstime Cor-
poration has made billions by importing and patenting scores of inventions from alternate time tracks. But the company's founder has
jumped from the balcony of his thirty-sixth-floor luxury apartment, the
latest in a series of inexplicable suicides that began only a month after
Crosstime started. A detective embarks on an investigation and gradually realizes the truth. People have committed suicide because of the
knowledge of the other versions of themselves, the might-have-beens
that lived lives that were less lonely or more fulfilled. They have committed suicide because they know that, if all possibilities happen, nothing they do ever matters; whatever decision they make, the opposite
decision will also be made in some other reality. They have committed
suicide out of despair. Finally, the knowledge becomes too much for
the detective as well, and he puts a gun to his head and fires. The hammer falls on an empty chamber. The gun jerks and blasts a hole in the

---

   * Although this is a possibility, the cause of most cancers is considered to be "multifac-
torial." Up to half a dozen factors such as genetic predisposition and exposure to carcino-
gens may be necessary in order to trigger a given cancer.

   † Larry Niven, "All the Myriad Ways," in *N-Space* (London: Orbit, 1992), 62.

ceiling. The bullet tears a furrow in his scalp. The bullet takes off the top of his head . . . and so on, ad infinitum.

We are fortunate indeed that decoherence ensures that we experience only one reality at a time. Granted, there are connections between the different realities. They make possible everything from semiconductors in ordinary computers to lasers to nuclear reactors. "But, luckily, our minds are adapted to experience only one reality out of the myriad realities," says Tegmark.

But having more than one reality has its benefits. For instance, it could solve a difficulty physicists have with time travel. Astonishingly, the laws of physics—specifically, Einstein's general theory of relativity—appear to permit it. The reason is straightforward. General relativity recognizes that time can flow at different rates for different observers. Time slows for an observer traveling very fast relative to another observer. It also slows for an observer experiencing strong gravity relative to an observer experiencing zero gravity. Because of these effects, it is possible to imagine two observers whose clocks tick at wildly different rates. While one observer lives from Monday to Friday, the other goes only from Monday to Tuesday. Now imagine a bridge, or tunnel, between the two observers. Such a tunnel is permitted to exist by general relativity; it is known as a "wormhole." By going down the wormhole, one observer can go back in time from Friday to Tuesday.

There are a few differences between this kind of "time machine" and the type described by science fiction writers such as H. G. Wells. For one thing, you have to travel through space to travel through time. And for another, you cannot go back to a time before your time machine was built. But the fact remains that time travel appears to be permitted by the laws of physics. This is unsettling for physicists because if time travel is possible, all sorts of paradoxes raise their ugly heads.

The concerns have prompted the English physicist Stephen Hawking to propose the "chronology protection conjecture," which in effect bans time travel. It is not proven, but in its support Hawking says: "Where are the tourists from the future?"

The most famous time-travel paradox is the "grandfather" paradox. A man goes back in time and shoots his grandfather. But if he shoots his grandfather, how can he ever be born to go back in time and do the dirty deed?

The Many Worlds hypothesis resolves the paradox simply. The man goes back and kills his grandfather. However, according to David

Deutsch, the grandfather who is killed is not the grandfather in the killer's reality; he is another version of the grandfather—one who resides in another reality in which the murderer is never born.

One of the reasons physicists originally disliked the Many Worlds idea—and some continue to dislike it—is that they have a horror of such a nightmarish proliferation of universes. This was, and is, an emotional response rather than a scientific one. However, physicists are happy to accept that we live in a universe with four space-time dimensions, the fourth dimension of which is beyond our powers of perception. Many physicists believe we live in a universe with ten space-time dimensions, with six "rolled up" too small to be perceived. If we have learned anything from science in the twentieth century, with its warped space-time and matter popping into existence out of empty space, it is that the underlying reality of the universe is nothing like the everyday reality of our senses. "Nature is under absolutely no obligation to make things easy for human brains or human senses," says Tegmark. "And that means it is under no obligation to provide us with a single reality rather than a bewildering infinity."

The whole reason for Everett's proposal of the Many Worlds idea was to reconcile the notions that big things such as tables are evidently localized in one place whereas the atoms that make up big things are not necessarily localized at all. This prompts an obvious question, one that has been conveniently skirted in all the discussion so far: why can atoms be in many places at once? The answer has to do with a peculiar fact. In countless experiments, particles such as atoms have been observed to have wavelike properties. For instance, they can bend around obstacles in exactly the same way as water or sound waves do.

It is a characteristic of all waves that they obey a "wave equation," which determines what kinds of waves can exist. Two of the many different waves permitted to exist by the wave equation for water waves are a big, rolling wave and a small ripple. Now—and this is the crucial issue—the wave equation also permits the existence of a wave that is a combination of the two waves. In that case, we are talking about a big, rolling wave with a small ripple superimposed on it.

This has earth-shattering implications for the matter of which we are made. For if atoms and their like are described by a wave equation—which they are; it is called the Schrödinger equation—then if there is a wave representing an atom over here and another wave representing a particle over there, it is perfectly possible for there to exist a wave that is a combination of the two waves. Such a "superposition" of waves represents an atom that is simultaneously over here and over there.

That particles like atoms behave like waves turns out to be behind all the weirdness of quantum theory, or wave mechanics. To reflect the wave nature of a quantum system such as an atom, physicists describe it by an entity called a "wave function." The wave function is just a mathematical convenience, a device for carrying out calculations. At least that is what everyone has believed since the birth of quantum theory. However, what everyone has believed may be wrong, according to an English physicist based in the United States. If Humphrey Maris is right, the wave function is no mathematical convenience. Far from it. It's a glimpse of the ultimate reality that underpins the world. We are all ultimately made from wave functions.

3

Dividing the Indivisible

A claim that the basic building blocks of matter can be split could have profound implications for the nature of ultimate reality

Meditating about the hidden nature of things, the Greek philosopher Democritus came to the problem of the structure of matter and was faced with the question of whether or not it can exist in infinitely small portions. On the basis of some obscure philosophical considerations, he finally came to the conclusion that it is "unthinkable" that matter could be divided into smaller and smaller parts without any limit, and one must assume the existence of "the smallest particles which cannot be divided any more." He called such particles "atoms."

—George Gamow, *Mr. Tompkins in Wonderland*

U p on stage, the magician's female assistant lies down in a coffin-shaped box. The magician closes the lid and, to the gasps of the audience, begins to saw the box in half. But this is no sadistic joke; this is entertainment. In their heart of hearts, all the audience members know that, when the magician opens the lid again, out will step his assistant completely unscathed. What a terrible shock it is, then, when the lid folds back and out step . . . two perfect half-size copies of the magician's assistant.

Sound like nonsense? Well, something close to this may actually be possible, according to Humphrey Maris of Brown University—not in the everyday world, but in the microscopic world of atoms and their constituents. "It seems that an 'electron' can split into two fragments which behave to all intents and purposes like half-electrons," says Maris.

Consider what Maris is saying. The electron is the lightest particle in the atom, and the one with the greatest claim to being absolutely fundamental. In fact, in the 104 years since the electron's discovery by J. J. Thomson at Cambridge, not one experiment has revealed the slightest indication that the electron is anything but indivisible, the modern-day incarnation of Democritus's "uncuttable" atom.* The claim that electrons are cuttable is therefore nothing short of a bombshell dropped into the heart of physics.

But this still fails to convey the true significance of Maris's claim; 104 years of experiments having failed to show the electron as anything but indivisible is no proof that it really is indivisible. It is perfectly conceivable that tomorrow or next year or in ten years' time an experiment could reveal the electron to be built from even smaller things. Although this would surprise physicists, it would not necessarily undermine their whole field nor merit a mention on the evening news. No, what Maris claims is not simply that the electron is assembled from microscopic Lego blocks, but that the electron can be split apart regardless of whether it is made of smaller building blocks.

The Weird World of Matter Waves

How can something be cut into smaller pieces if it is not made of smaller pieces? Exactly. Now we can begin to appreciate the extraordinary nature of Maris's claim. Ultimately, this isn't about whether the electron is made of tinier bits—"subelectrons," if you like—but about the nature of the fundamental reality that lies beneath the surface of our familiar, everyday world.

All ordinary matter is ultimately constructed from microscopic building blocks—atoms and their constituent particles. This view was modified slightly in the 1920s when physicists discovered that the building blocks of matter behave in peculiar, counterintuitive ways. For instance, particles such as electrons can bend around obstacles in their path in much the same way that the sound waves from someone shouting can bend around the corner of a house.

---

* Democritus was a Greek philosopher who lived in the fifth century B.C. He is famous for his idea that matter is ultimately made of tiny grains that cannot be cut into smaller pieces. Because the Greek phrase for "uncuttable" is "a-tomos," Democritus christened the grains "atoms."

One of the consequences of this wavelike behavior of subatomic particles is that they are not necessarily localized at one point in space, like microscopic billiard balls. Instead, they possess something of the character of a spread-out water wave. Before being "observed," a particle has a certain probability of being found at one location, another probability of being found at another location, and so on.

To describe the wavelike properties of fundamental particles, physicists invented the "wave function." This encapsulates all that is knowable about a particle. The wave function spreads out with time rather like a water wave rippling across the surface of a pool, and the manner in which it spreads out is governed by an equation discovered by the Austrian physicist Erwin Schrödinger in the 1920s and known ever since as the Schrödinger equation. To find out where a particle is most likely to be at any future time, physicists simply use the Schrödinger equation to predict how the particle's wave function will develop, extract from the wave function at the later time the most probable location of the particle, and then discard the wave function.

The wave function is nothing more than a convenient device for calculation. It cannot be observed directly, as a water wave can. It isn't really a "thing" at all; it's a wave of mathematical probability. This, at least, has been the received wisdom. Until now.

Maris's claim changes everything. What can be cut in half, he says, is the electron's wave function. And, because he further claims that the result is two half-electrons, the inescapable conclusion is that the wave function of an electron is the electron. In other words, what everyone, for the last eighty years, has considered a mere mathematical device is the ultimate reality that underpins our world. You and I and everyone are ultimately made of wave functions.

Splitting the Wave Function

Maris's area of expertise is matter at ultralow temperatures. The possibility that an electron wave function might split in two first occurred to him while he was thinking about liquid helium, a substance that liquefies only below -269 degrees Celsius, or 2.17 degrees above absolute zero.*

---

*   Absolute zero is the lowest temperature attainable. When an object is cooled, its atoms move more and more sluggishly. At absolute zero (which on the Celsius scale is equal to -273.15 °C), they stop moving altogether.

Liquid helium is arguably the weirdest substance known to humankind. At temperatures below 2.17 degrees, it becomes a "superfluid" that can flow without friction, squeeze through impossibly small holes, and, most remarkable of all, run up hills. Maris, however, was interested in another peculiar feature of the liquid: "electron bubbles."

Electron bubbles were first made in the late 1950s when physicists squirted high-speed electrons into liquid helium. Such electrons gradually slow down and come to a halt in the liquid. However, this presents them with a problem. If at all possible, an electron likes to circle the central "nucleus" of an atom much the way Earth orbits the Sun. In liquid helium, however, all the atoms possess their full complement of electrons. With no room at the inn, an interloper has no choice but to lodge itself in the space between helium atoms.

The electronic outcast contributes to its own isolation because it fiercely repels the helium atoms all around it.* After pushing them away, it ends up floating in an empty bubble in the liquid. This electron bubble, though too small to be seen with the naked eye, is vast on the atomic scale, a cavity from which almost a thousand helium atoms have been rudely ejected.

The electron in the electron bubble, like an electron anywhere else, is described by a wave function. However, there is a restriction: only certain wave functions can survive within the bubble. Imagine a water wave sloshing about inside a hollow container. Most water waves will bounce around chaotically within the cavity, changing from moment to moment and quickly dying away. However, a small handful of possible waves will be "tuned" to the shape of the cavity; the waves bouncing off the walls will unite with the waves in the interior of the cavity to make a combined wave that will not die away. The forces of creation and destruction will be perfectly balanced everywhere, creating a wave that appears to stand still.

The set of permitted wave functions for the electron in the electron bubble contains just these "standing waves." Each, it turns out, corresponds to an electron with a different energy. The most likely state in which to find the electron is the one with the least energy, because this is the least energy-hungry. In that case, the wave function is "spherically symmetric," which means that the height, or amplitude, of the

---

* Electric charge comes in two forms—positive and negative. Like charges repel, whereas unlike charges attract. Because all electrons have the same electrical charge, they repel one another, and this includes electrons lodged between helium atoms and electrons orbiting inside helium atoms.

wave function falls off in the same way in all directions. Because the probability of the electron being at any location in space is determined by the wave function, the electron in the lowest-energy state is equally likely to be found in any direction.*

Although the lowest-energy state is the most likely state, the electron in the bubble need not be in that state. If the electron has a surplus of energy—and in an experiment this can be supplied by laser light shone on the bubble—it can be in the next-highest energy state. It turns out that the wave function in this state, known as the first "excited" state, is not spherically symmetric but dumbbell shaped. This means that the most probable places to find the electron are on opposite sides of the bubble, rather like the North and South Poles of Earth.

If an electron spends more time near the north and south poles of its bubble, then loosely speaking it can be thought of as a trapped bee preferentially hammering away at the north and south poles of its prison. Maris realized that this would tend to elongate the bubble in the north and south directions. In 1996, he sat down to calculate the precise effect.

The first thing he grasped was that the speed at which the bubble elongated depended on several factors. The most important was the thickness, or viscosity, of the liquid helium. If the liquid were relatively viscous, it would behave just like molasses, resisting the stretching of the bubble so that it would change shape smoothly and gradually from a sphere to a dumbbell.

Temperature was crucial to the process. Below 2.17 degrees above absolute zero, helium stops being a single liquid and starts being two wildly different liquids. One is the so-called normal phase and the other is the superfluid phase. Among the many bizarre things the superfluid can do is run forever without any friction slowing it. In other words, it behaves like an ultrarunny liquid with no viscosity whatsoever.

The two liquids—the normal and the superfluid—intermingle. At 2.17 degrees above absolute zero, when the superfluid first puts in an appearance, it is a negligibly small component of the overall liquid. However, as the temperature of the liquid drops below this threshold, the superfluid becomes more and more important. And, because it has absolutely no viscosity, its increasing dominance causes the viscosity of the liquid as a whole to plummet. Simply by lowering the tem-

---

* The probability is determined by the square of the height of the wave function.

perature of the liquid sufficiently, it is possible to achieve an arbitrarily tiny viscosity.

Maris realized that this had important implications for a stretching bubble. Liquid helium would indeed resist the elongation of an electron bubble, causing it to change slowly and smoothly from a spherical to a dumbbell shape. But only above a certain temperature. Below this "critical" temperature, which Maris calculated to be 1.7 degrees above absolute zero, the liquid was too runny. As the trapped electron hammered away inside, the bubble would stretch so quickly that it would actually overshoot the dumbbell shape. "The bubble would stretch into a long sausage, then it would develop a thin neck," says Maris. "The neck would get thinner and thinner until, finally, the bubble split into two smaller bubbles."

This splitting of a bubble into two "daughter" bubbles might appear a mundane and inconsequential event. After all, the bubbles in the soapsuds used for washing dishes split all the time and nobody gives this process a thought. However, the splitting of an electron bubble was an extraordinary event, without precedent in the world of physics. The reason? "Because if the electron wave function was dumbbell shaped and the bubble split into two, trapped in each of the daughter bubbles would be half an electron wave function," says Maris.

The implications were staggering. If, as quantum theory indicates, the wave function is the essence of the electron, cleaving the electron wave function would actually break the electron in two. Splitting a bubble in liquid helium was as far from mundane as it was possible to imagine. It was the means to divide the indivisible.

This was revolutionary, explosive stuff. If Maris were going to convince anyone else, he would need to be sure of his ground. First, he needed to check his calculations. Then he needed to carry out some experiments to make sure nature did indeed do what his calculations indicated. His first port of call was the library at his university. It was there, flicking through back issues of physics journals, that he discovered something amazing. Some of the experiments he had in mind had already been done.

Solving a Thirty-Year-Old Puzzle

In the late 1960s, two physicists at the University of Minnesota had set out to measure the speed of electron bubbles traveling through liquid helium. Jan Northby and Mike Sanders made their electron bubbles move with an electric force field that they set up across the liquid. The moving bubbles, with their electrons inside, constituted an electrical current, because an electrical current is just a flow of electrically charged particles. By measuring the current—a relatively simple procedure—the two physicists were able to deduce the speed of their electron bubbles.

Northby and Sanders then shone a laser light on their liquid. They expected that this would give the electrons in the bubbles such a boost in energy that they would be kicked clean out of the bubbles. Because free electrons are much tinier and easier to shift than relatively large electron bubbles, the electric force field ought to whiz them through the liquid, causing the electric current to leap in size. This is precisely what Northby and Sanders observed in their experiment.

With the theory so neatly confirmed, this might have been the end of the matter. However, it wasn't. The whole idea behind the experiment turned out to be flawed. When an electron was knocked out of its bubble prison, it did not remain free for long. Instead, it repelled all the helium atoms in its vicinity and promptly formed another electron bubble. Because a new electron bubble was created for each one destroyed, shining a light on the liquid helium should have had no effect at all on the electrical current flowing through the liquid. So why did it? For decades no one could figure it out—until Maris came along. "If electron bubbles could split in two, the thirty-year-old mystery might be solved," he says.

According to Maris, the laser light used by Northby and Sanders did not kick the electrons from electron bubbles. "It merely boosted them from the lowest energy state to the first excited state—the one with the dumbbell-shaped wave function," he says.

Crucially, Maris noticed that the two physicists had carried out their experiment with the liquid helium below 1.7 degrees above absolute zero. This was the critical temperature below which nothing could have stopped the electron bubbles from splitting. "The splitting of bubbles would have created more bubbles," says Maris. "Being smaller, they would move more quickly through the liquid, boosting the electrical current, just as was observed."

A similar experiment was carried out in the early 1990s by two physicists at AT&T's Bell Labs in New Jersey. They too found that electron bubbles, contrary to expectations, moved faster when light was shone on them.

And these were not the only inexplicable experiments that Maris discovered as he scoured the journals. Other physicists had found a more precise way of studying how quickly electron bubbles moved through liquid helium. They included Gary Ihas and Mike Sanders at the University of Michigan in 1971 and Van Eden and Peter McClintock at the University of Lancaster in 1984. These physicists created a short burst of about a million electron bubbles and then carefully timed the burst as it was propelled by an electric force field across a container of liquid helium.

Because the electron bubbles in the burst were all created together, the experimenters expected to see all the electron bubbles cross the finish line together. Imagine their consternation when the bubbles arrived at three distinct times.

According to Maris, this mystery too is solved if electron bubbles split into two. In all the experiments the electron bubbles were created by injecting electrons into the liquid from an electrical discharge—the laboratory equivalent of lightning. "An unavoidable by-product is light," says Maris. "This light could have boosted electrons in electron bubbles from the lowest-energy state to the first excited state so that the bubbles split."

According to Maris, some of the bubbles could have split into more than two daughter bubbles. Or some of the daughter bubbles could have divided again. Whatever the precise details, the result would have been a whole range of different-sized bubbles, containing a whole range of different "fractional electrons." No wonder they arrived at the finish line at a range of different times.

After his trip to the library, Maris now had far more than he could have hoped for: a whole series of baffling experiments, dating back to the late 1960s, all of which met the conditions necessary for bubble splitting. All along, his hunch had been that if an electron wave function split, so too would the electron itself. Here, it seemed, was the evidence. The baffling experiments suddenly made sense if electrons with a fraction of their normal charge were on the loose.

Maris was now more confident of his idea. But it was heresy and he did not dare announce it yet. Instead, over the following years, he checked and rechecked his calculations. "It took time to get used to the idea and pluck up the courage to tell the physics community," he says. "But finally, in June 2000, I decided to go public."

The Announcement

The venue was a conference in Minneapolis. Usually, people are given about thirty minutes to speak at such conferences. However, such was the perceived importance of Maris's work that, in addition to the standard slot, he was given two hours on the Saturday night before the conference. More than a hundred physicists crowded the auditorium where he spoke and, when he finished presenting his evidence for the splitting of electrons, the questions came thick and fast. "What I was saying created quite a stir, as I knew it would," he says.

"My first reaction was extreme skepticism, like everyone else," says McClintock. "However, nobody else has come up with a plausible explanation of the baffling electron bubble experiments. If electrons can indeed split into fragments, it offers a possible way out."

"I was very nervous that someone would find a hole," admits Maris.

"There were lots and lots of questions but Humphrey had an answer for everything," says McClintock. "He'd obviously thought long and hard about the whole thing."

"I have never seen any scientist defend a theory so deftly and as gracefully as Humphrey Maris," says Ben Stein of the Institute of Physics in College Park, Maryland. "To every criticism he provided a plausible counterargument."

"To my relief nobody dismissed the idea out of hand," says Maris.

"If Humphrey is correct, it means a Nobel Prize," says Gary Ihas, now at the University of Florida.

"However, the odds are very high that it is not correct," says Nobel Prize winner Philip Anderson of Princeton.

Anderson's sentiments are shared by theorists. "The idea of a wave function splitting in two is totally incompatible with quantum theory," says Anthony Leggett of the University of Illinois at Urbana-Champaign. Leggett admits that there could be something wrong with quantum theory. "However, given its overwhelming success in explaining the world, this is highly unlikely," he says.

Like Leggett, most physicists are convinced that Maris's claim is tripped up at the first fence. "However, it's not at all obvious why," says Anderson.

But, say, just for a moment, that electrons really can split into two. What would it mean?

What Does It Mean?

For one thing, it would totally change our idea of the wave function. After all, if the wave function of an electron can be split and the result is two half-electrons, then it is hard to avoid the conclusion that the wave function is the electron. It isn't a mere mathematical convenience, as physicists have believed for nearly eighty years. It is the ultimate reality that lies beneath the surface of the world. You and I are ultimately made of wave functions.

Many questions spring to mind. What exactly is a half-electron like? Does it have half the normal charge? Does it have half the mass of a standard electron? And what are its other properties? Maris admits to being as perplexed as everyone else. "At this time, I simply don't know," he says.

Particles such as electrons are slippery, ill-defined objects that have a certain probability of being over here, a certain probability of being over there, and so on. Only when they are observed are they forced to make up their minds. What then would happen if you cornered and probed an electron bubble with half an electron wave function in it? "I believe you will see a complete electron," says Maris. "But then what happens to the electron bubble with the other half of the electron wave function in it? Does the bubble suddenly find itself empty and implode instantaneously? Perhaps it does. I don't know."

Maris's hypothesis seems to throw what we know about the world into confusion. Could a half-electron form naturally? The existence of such fractionally charged particles would appear to undermine the whole idea of there being indivisible building blocks out of which the universe is made, which is a foundation stone of physics. However, it undermines physics only if half-electrons are common. And Maris says he can see no circumstances apart from those of a contrived experiment with ultracold liquid helium in which fractional electrons can form.

Does an electron bubble have to have liquid around it? If it does not, what about the possibility of taking a bubble out of the liquid? Could you attach the electron fragment to an atom? The structure and function of the material world, including our bodies, depend on the chemical reactions between the ninety-two naturally occurring species of atom. Those reactions in turn depend on the number and arrangement of atomic electrons. Whole electrons. What would chemistry be like with fractional electrons? Neither Maris nor anyone else has the faintest idea.

Maris, having lobbed his bombshell into the world of physics, seems happy to sit on the sidelines smiling quietly to himself at the chaos. "The questions are only going to be answered by more experiments," says McClintock.

Maris and many others are currently doing those experiments. "Already, the results are encouraging," he says. "What I have come up with is an intriguing puzzle. I want people to think. In fact, I would be happy if I was completely wrong but made a lot of people think."

Maris's claim remains controversial. Most people still think of ultimate reality as built of particles like tiny billiard balls rather than of slippery wave functions. But this still leaves a nagging question: what is a particle? It's so basic a question that few physicists ever bother to ask it. However, one British physicist has. His name is Mark Hadley, and the answer he has come up with may knock your socks off. Hadley maintains that fundamental particles—the building blocks of you and me—are nothing less than tiny time machines.

4

All the World's a
Time Machine

*The two great theories of twentieth-century physics
might at last be united—if atoms contain time machines*

When one reaches the country of the fourth dimension, when one
is freed forever from the notions of space and time, it is with this
intelligence that one thinks and one reflects. Thanks to it, one
finds himself blended with the entire universe, with so-called future events, as with so-called past events.

—Gaston Pawlowski,
Voyage to the Country of the Fourth Dimension

Time is nature's way of keeping everything from happening at
once.

—Graffito, men's room, Austin, Texas

Our world is built from tiny, indivisible grains of matter; the idea
originated with Democritus and was finally confirmed in the twentieth century. At one time it was thought that nature's indivisible grains
of matter were atoms. Nowadays, they are believed to be quarks and
leptons, objects quite a bit smaller than atoms.* But what exactly are
the ultimate building blocks of matter?

One person who thinks he knows the answer is the British physicist
Mark Hadley. "Nature's fundamental particles are tiny loops of time,"
he says. "What you and I are ultimately made of is time machines!"

Surprisingly, few physicists ever think to ask the obvious question:
what is a particle? One physicist who did, however, was Albert Einstein.

---

* Quarks are the constituent parts of the protons and neutrons that form the central
nuclei of atoms. The most common leptons are the electrons that orbit nuclei.

Throughout his life, Einstein struggled to understand the universe at its deepest, most fundamental level, and this necessarily required penetrating the secret of fundamental particles. He believed he had found a guiding light in his theory of gravity—the general theory of relativity.

According to general relativity, the presence of matter distorts, or warps, the fabric of space-time. It is this distortion that produces the effects we attribute to the force of gravity. The warpage of space-time, in short, is gravity. Although the Sun appears to exert a force of attraction on Earth, trapping it in perpetual orbit, in reality there is no such force. Instead, the mass of the Sun warps the surrounding space-time into a valleylike depression, and Earth rolls around and around the rim of the valley much as would a marble in a bowl. We do not "see" the valley because it is a valley in four-dimensional space-time, something as impossible for us to perceive as our three-dimensional world would be to a creature that lived only on a two-dimensional sheet of paper.

At first sight, there would seem to be no conceivable connection between a theory that describes the motion of bodies such as planets and one that sheds light on the ultimate nature of fundamental particles. However, fundamental particles, by definition, are the simplest objects in the universe. It follows, therefore, that they ought to be built from the simplest possible building materials. And building materials, Einstein realized, come no simpler than space and time—the very raw material of general relativity.

Einstein's theory describes how space and time are distorted by the presence of matter. However, there is nothing in the theory that says space-time cannot have an intrinsic warpage, imposed in some as yet unknown manner.* This at least was Einstein's tack. Perhaps general relativity permitted the existence of some kind of microscopic buckle in space—one that behaved exactly like a tiny object with a mass. Alas, Einstein was destined to be disappointed. Despite his best efforts, he was unable to find a localized whorl in space that mimicked a fundamental particle such as an electron.

After Einstein's death, however, others took up the baton, including in the 1960s Charles Misner and John Wheeler, the man who invented

---

*   The space-time of our universe is intrinsically warped. This warpage was imposed at the birth of the universe during the explosion of the Big Bang. Because the warpage is in time as well as space, there is the possibility of the universe curving back on itself in time. This corresponds to it one day recollapsing back down to a Big Crunch, a sort of mirror image of the Big Bang. At present, no one knows for sure whether this will happen.

the term "black hole." But, like their illustrious predecessor, these physicists too hit a brick wall.

Enter Mark Hadley. Soon after asking himself the question—What is a particle?—he came across the work of Einstein, Wheeler, and Misner. He became convinced that they were on the right track. However, one peculiar thing struck him about their approach. In their quest to find something in general relativity that behaved like a fundamental particle, they had looked for a localized distortion not in the fabric of space-time but merely in space.

It seemed a surprising oversight. After all, general relativity describes how matter warps not just space but space-time. Space-time is an amalgam of space and time; the two are merely different faces of the same coin. This is far from obvious to us in everyday life. However, Einstein realized that if we could travel extremely fast—close to the three-hundred-thousand-kilometers-per-second speed of light—or experience ultrastrong gravity, an obscuring veil would be lifted from our eyes. Suddenly, we would see the unity of space and time. In light of this, it made sense to view our universe as having not three dimensions of space and one of time but four dimensions of space-time.*

So why did Einstein, and then Wheeler and Misner, ignore the message in general relativity and try to explain fundamental particles as mere buckles in space? For good reason. A tiny piece of space-time that could appear as localized as a microscopic particle would have to be grossly warped. Grossly warped space-time necessarily contains grossly warped space and grossly warped time. Grossly warped space was not a particular worry to Einstein and the others. However, the idea of grossly warped time made them break out in a cold sweat.

Conceivably, time might be so badly warped that it bent back on itself. If this ever happened in the everyday world, bizarre things could happen. You might feel raindrops on your face before it started raining. You might arrive at work before you left home. You might die before you were born. Such events are said by physicists to violate the principle of "causality"— the idea that a cause must always precede an effect. This is one of the most cherished ideas in the whole of physics. Without causality, and the

---

* Oddly enough, the first person to realize this truth was Hermann Minkowski, Einstein's former mathematics professor at Zurich's Federal Polytechnic School. Minkowski had famously dismissed his student as a "lazy dog," the scientific equivalent of turning down the Beatles. To his eternal credit, however, Minkowski later recognized his mistake and became an evangelist for his student's revolutionary ideas.

order it imposes on our experience of the world, there would be total chaos. The entire edifice of physics would come tumbling down. At least, this was what Einstein, Wheeler, and Misner feared.

Hadley, however, was not convinced things would be so bad. For all his greatness, Einstein had been wrong about two major predictions of his general theory of relativity. He had refused to countenance the idea of black holes, regions of space where gravity holds light itself prisoner, and he had not believed that the universe could be expanding in the aftermath of an explosion. Could he be wrong about a third prediction—that the violation of causality would wreck physics and so could never be permitted to happen?

It seemed to Hadley that Einstein and the rest had seen the road stretching ahead but balked at taking it. He decided to go all out. Fully embracing the spirit of general relativity, he decided to explore the idea that fundamental particles are localized distortions not simply in space but in space-time.

Hadley was working in industry at the time, having left the University of Warwick with an undergraduate degree in physics in 1979. Recognizing that his mathematics was not up to handling the formidable apparatus of general relativity, he went back to Warwick and embarked on a master's degree. He followed it up with a Ph.D. "To make real advances in science, you always have to go out on a limb," says Gerard Hyland, Hadley's thesis supervisor at Warwick. "By adopting such a strategy, Mark took a very big risk."

Rather than seeing causality violation as a nightmare, Hadley viewed it as a virtue. It occurred to him that it might help solve one of the outstanding problems of modern physics: how to mesh general relativity and quantum theory.

Why Unite Relativity and Quantum Theory?

General relativity is a theory about gravity. However, the gravitational attraction between small things such as atoms, or even people, is far too small to notice. The force becomes appreciable only for large accumulations of matter such as planets, stars, galaxies, and the universe as a whole. In essence, therefore, general relativity is a theory of the very large. Quantum theory, on the other hand, describes the submicroscopic world of atoms and their constituents. It is a theory of the very small.*

Because general relativity and quantum theory relate to domains that differ widely in size, and that do not seem to overlap, there would appear to be no pressing need to unite them into a single overarching theory. Appearances, however, are deceptive. Although the universe is undeniably big at present, it has not always been so. Astronomers observe that the universe is expanding, its constituent galaxies flying apart like pieces of cosmic shrapnel. From this, they conclude that it must have been smaller in the past. If we imagine the history of the universe running backward like a movie in reverse, we come to a time about twelve to fourteen billion years ago when the universe was smaller than an atom. This was in the immediate aftermath of the Big Bang explosion in which the universe was born. It follows that, if we are ever to understand the origin of the universe, we will first have to unite the theory that describes the very large — general relativity — with the theory that describes the very small — quantum theory.

Unfortunately, there are huge difficulties in doing this. At a fundamental level, the two theories appear to be utterly incompatible.

Incompatible Theories

Quantum theory was born of the struggle to reconcile two apparently irreconcilable features of nature — light and matter. By the early twentieth century, it was clear to many physicists that matter was ultimately made out of impossibly tiny particles — atoms. Light, on the other hand, appeared to be made of waves, reminiscent of the ripples that spread out on a pond after the impact of a raindrop. There were no problems with either of these pictures — until physicists began to probe the interface where light meets matter.

This interface is of fundamental importance to the everyday world. If atoms in the heated filament of a bulb did not emit light, we would

---

* To be precise, quantum theory is a theory about entities that are isolated from their surroundings. Although in principle big objects can be isolated, in practice an object's isolation is broken if only a single photon of light from the outside world bounces off it. Because it is much easier for small objects such as atoms to maintain their isolation than it is for big objects such as people, quantum theory is largely a theory of the very small.

not have instant illumination in our homes. If atoms in the retina of your eye did not absorb light, you would not be able to read these words. However, the absorption and emission of light by atoms poses a conundrum. An atom is localized, confined to a tiny region of space, whereas a light wave is decidedly not localized; it is spread throughout a comparatively large volume of space. When light is absorbed, how does it squeeze down to the size of an atom? When an atom emits light, how does it manage to cough out something so big? Clearly, there is something wrong with our picture of atoms, or light, or both.

Common sense says that the only way light can be absorbed or emitted by a small, localized atom is if it too is small and localized. The extraordinary picture that emerged in the first decades of the twentieth century was that light must be both a spread-out wave and a localized particle. It was a weird entity that could behave not only as a ripple on a pond but also as a stream of particles, "atoms of light," which were later dubbed "photons."

The new, quantum worldview was weirder even than this. It turned out that there was complete symmetry between light and matter. Not only could light behave like particles, but particles could behave like waves. To the dismay of many physicists, the constituent parts of matter turned out to be slippery and elusive, resisting all attempts to pigeon-hole them. Sometimes they manifested themselves as waves, sometimes as particles. However, they were neither waves nor particles but something else for which there was no analogue in the everyday world and, consequently, no word in our language.

But the new and revolutionary picture of light and matter was not simply weird. As Einstein was the first to realize, it was catastrophic for accepted physics, totally incompatible with everything that had gone before.

Look closely at a window; you will see a faint image of yourself. Glass is not perfectly transparent. It transmits only about 95 percent of the light that strikes it, and reflects the rest. This is perfectly easy to understand if light is a wave. The wave simply splits into a big wave that carries on and a much smaller wave that doubles back on itself. You can see something similar when the bow wave from a speedboat encounters a half-submerged piece of driftwood. However, what is straightforward if light is a wave is extremely difficult to understand if light is a stream of identical, bulletlike photons.

The problem is this. If all photons are identical, then its stands to reason that, on encountering a window, each will be affected in an identical way. (We are talking about a uniform window whose thickness and transparency are the same everywhere.) How then is it possible that 95 percent of the photons are transmitted through the glass and 5 percent are reflected? Clearly, it isn't possible — not unless we are willing to redefine the word "identical."

This was the desperate remedy physicists were forced to accept in the 1920s. For photons, "identical" does not mean what it does in the everyday world. It has a diminished meaning. "Identical" means having an identical chance, or probability, of doing something. Each of the photons striking the window has a 95 percent chance of being transmitted and a 5 percent chance of being reflected. There is no predicting whether a given photon will be transmitted or reflected. Ultimately, everything comes down to chance. And what is true of photons is true of all microscopic particles. What they get up to is inherently unpredictable. The shocking truth is that the whole universe is founded on random chance.

This conclusion is absolutely unavoidable the moment we accept that light and matter come in discrete chunks, or "quanta." To Einstein, quantum theory was a disaster for physics. Until his dying day, he declared: "I shall never believe that God plays dice with the world." Unfortunately, as Stephen Hawking has pointed out: "Not only does God play dice, he throws them where we cannot see them!"

Here then is the fundamental incompatibility between quantum theory and general relativity. General relativity, like every theory of physics before it, is a recipe for predicting the future. If a planet is here now, in a day's time it will have moved over there, by following this path. All these things are predicted by the theory with absolute certainty. Compare this with quantum theory. For an atom flying through space, all we can predict is its probable final position, its probable path. The very foundation stones of general relativity, such as the trajectory of a body through space, according to quantum theory, are a fiction.

The task faced by physicists is therefore to unite a theory that deals with probabilities with one that deals with certainties. To say this is a challenge is a bit of an understatement.

Which Is More Fundamental?

General relativity and quantum theory cannot both be correct. If we could somehow go back to the first few moments after the Big Bang, when the whole universe was squeezed into a volume far smaller than an atom, we would see one of the two theories break down. All the bets are that that theory would be general relativity.

The strongest reason for believing this is that general relativity appears to contain the seeds of its own destruction. It predicts that, if the expansion of the universe were to be run backward, the universe would get smaller and denser and hotter without limit. In other words, a point would eventually be reached when the universe was infinitely dense and infinitely hot. Technically, this is known as a "singularity." This particular singularity is a singularity in time. However, general relativity also permits singularities in space. These occur when gravity triggers the runaway shrinkage of a star to form a black hole. Here, the theory predicts the star getting denser and hotter without limit.

Singularities, in which things such as temperature skyrocket to infinity, are physical impossibilities. They are a sure sign that a theory is being stretched into a domain where it no longer has anything sensible to say. A similar singularity loomed large in the world of physics in the early part of the twentieth century.

Experiments by the New Zealand physicist Ernest Rutherford and others had revealed that an atom consisted of a tiny nucleus, which accounted for most of its mass, about which ultralight electrons orbited like fireflies around a campfire. The problem was that orbiting electrons should give out light, rapidly losing energy and spiraling down into the central nucleus. Within a mere millionth of a second they should have piled up in a point of infinitely dense electric charge—a singularity. Physics appeared to predict the nonexistence of atoms.

Disaster was averted by the efforts of a young Dane named Niels Bohr. Bohr took the idea of light coming in bulletlike chunks and fearlessly applied it to the atom. Orbiting electrons were not free to give out light of any energy. Instead, Bohr maintained, they could emit energy only in bundles of a certain size. These would later be known as photons. So severe was the restriction on the light that could be emitted that it was difficult for an electron to lose energy, and impossible for it to lose all of its energy.* Thus were electrons prevented from spiraling all the way into the nucleus; the atom was stabilized against catastrophe.

Quantum theory's success in staving off a singularity in the heart of atoms echoes to this day. It has encouraged the belief among physicists

---

*   The impossibility of an electron losing all of its energy and sinking to the heart of an atom is the result of a related quantum edict known as the Heisenberg uncertainty principle.

that quantum theory might also stave off the singularity in the Big Bang and the singularity at the heart of black holes. Almost everyone therefore subscribes to the view that quantum theory is a more basic theory than general relativity, and that eventually general relativity will be shown to emerge from quantum theory.

It would take a brave man to maintain anything different. Mark Hadley is such a man. He is turning accepted ideas on their head by claiming that quantum theory is actually a spin-off of Einstein's theory of gravity. While everyone else is struggling to discover how certainty can arise from uncertainty, Hadley believes he can see how uncertainty can arise from certainty. The key is the one thing that gave Einstein waking nightmares: causality violation.

Uncertainty from Certainty

The essence of Hadley's proposal is that a subatomic particle such as an electron is a tiny region of space-time so dramatically warped that it bends back on itself like a knot. Such a region necessarily contains a loop in time. "The time loop is the crucial ingredient which enables general relativity to reproduce the effects of quantum theory," says Hadley.

In a time loop, time loops back on itself in much the same way that space loops back on itself on an athletics track. The technical name physicists give such loops is "closed timelike curves." But, to the rest of us, they have a more common name: time machines. Hadley's contention is that fundamental particles, and therefore you and I, are ultimately made of time machines.

In the everyday world, time flows one way, remorselessly into the future. However, from the point of view of a time machine, time is a two-way street, with past and future equally accessible.* Crucially, a fundamental particle containing a time loop can be affected not only

---

* The reason time machines, or causality violation, are features of general relativity is not difficult to understand. For observers experiencing different gravity or traveling at different speeds, time flows at different rates. For instance, time flows more slowly for someone in strong gravity (which means we age slightly faster at the top of a tall building than at street level). Imagine that two people—one in a strong gravitational field and one in a weak gravitational field—synchronize their watches on Monday. When the one in the weak field gets to Friday, the one in the strong field may still have reached only Tuesday. If you create some kind of bridge from one observer to the other, it would be possible to go back in time from Friday to Tuesday. In theory, such bridges could exist. They are known as "wormholes."

by events in its past but also by events in its future. "Not surprisingly, this changes everything," says Hadley.

Pause for a moment to think what it means. The only events that affect you are events in your past. You wake up with a cold today because someone sneezed near you yesterday. You are alive today because your parents bumped into each other and hit it off before you were born. If, in addition, you were also affected by events in the future, this would make a profound difference to your life.

Imagine that your state today—whether, say, you are alive to read these words or not—is determined by events that happen tomorrow. Perhaps a bus will run you over and kill you. Or perhaps it won't. In this particular example, the future events that determine your state today are mutually incompatible. It is impossible to be simultaneously killed and not killed. Consequently, your state today will necessarily be ill defined. Are you reading these words or not? If it depends on future events, it may be impossible to be 100 percent sure.

In the same way, if the properties of a fundamental particle—say, where exactly it is located—are determined by measurements that can be made in the future, the location of the particle may also be ill defined. It will not be possible to say it is over here, or that it will do a particular thing, with 100 percent certainty—only that it has a certain probability of being over here, a certain probability of doing that particular thing. Thus, miraculously, time loops can conjure the uncertainty of quantum theory from the certainty of general relativity. This raises the hope of reconciling the irreconcilable.

Hadley has another way of looking at it. "Imagine a blind man throwing a ball into a wastepaper basket," he says. "Getting the ball to the location of the basket depends only on releasing the ball in just the right direction with just the right speed. Moving the basket makes no difference to the path it takes through the air. However, imagine jiggling a rope instead. The shape of the undulating wave that travels along the rope depends not only on what is happening at the end you are holding but also on what is happening at the other end. Whether, for instance, the rope is free or tethered in some way."

Physicists call constraints at either end of some interval "boundary conditions." "A quantum particle has much in common with the rope," says Hadley. "There is another end—an unknown boundary condition in the future. Consequently, not everything is determined."

The unknown future boundary condition, according to Hadley, is why quantum events never occur with certainty, only with a particular probability. "Probabilities are not fundamental but exist because some of the boundary conditions are undetermined," says Hadley. "It's like not being able to predict whether a coin will fall heads-up or tails-up because of not knowing the exact initial conditions—how fast and in what direction the coin is tossed, the effect of drafts, and so on."

So much for the inability to know for sure where a particle is. Can time loops explain anything else? According to Hadley, the answer is yes. Take "nonlocality."

Fundamental particles that are born together—"created in the same quantum state," to be precise—forever after share a ghostly bond. Like identical twins, they seem to know about each other even when far apart. Consider two particles with oppositely directed "spins." If two particles are created whose spins point in opposite directions, their spins will always point in opposite directions. In other words, if one particle were somehow to be taken to the far side of the Moon—or the far side of the universe, for that matter—and flipped so that its spin pointed the other way, the particle left at home would flip too.

The incredible thing is that the particle back on Earth would react instantaneously—that is, as if the news of its partner flipping had traveled home at an infinite speed. However, as Einstein was the first to realize, no material object or signal of any kind can travel faster than the speed of light. The speed of light is enormous—a light beam can travel from the Moon to Earth in just over a second—but it is not infinite.

Everything in the universe is in some weird sense linked, because all particles—those out of which you are made and those that constitute the most distant galaxy—were once together in the same state during the Big Bang. The "spooky" connectedness of particles, in violation of the cosmic speed limit, was cited by Einstein as a reason why quantum theory had to be wrong. Unfortunately for Einstein, careful experiments carried out in laboratories since the early 1980s have confirmed that particles can indeed communicate with each other instantaneously.

Despite this, quantum theory remains compatible with the existence of an ultimate speed limit—the speed of light. This is because only limited kinds of information can be transferred instantaneously between particles. It would not be possible, for instance, to send a meaningful message.

The experiments have shown that quantum theory is right and Einstein wrong. However, this still leaves physicists with the problem of explaining how particles communicate with each other, apparently faster than light.

The problem goes away entirely, according to Hadley, if fundamental particles contain time loops. Because past, present, and future are all the same to them, in a sense fundamental particles live outside time. There is therefore nothing to stop such a particle reacting to an event before it actually happens. According to Hadley, this is all that is happening when a particle reacts instantaneously to its partner flipping direction. It is simply reacting before news of the event arrives.

Temporal Leakage

But if the atomic building blocks of matter contain time machines, why do they not play havoc with the everyday world? Future events do not appear to affect us. We do not arrive at work before we leave home. We do not die before we are born. And nobody, apart from Dr. Who, has managed to construct a working time machine.

Hadley concedes that this is a serious difficulty with his idea. If the potentially disastrous effects of time loops are not to leak out, they must somehow be cloaked from the world in which we live. But how? "In the world of physics there is only one such impenetrable cloak," says Hadley. "An event horizon."

An event horizon is the surface that surrounds a black hole. Think of it as a one-way membrane. Once a body plunges through the membrane, it can never escape the tremendous gravity of the black hole. Nothing inside, not even light, can ever get out and affect the world outside. The event horizon effectively seals the interior of a black hole from the outside universe.

Hadley therefore speculates that time loops are safely hidden behind event horizons. Time loops are believed to exist inside large-scale black holes that are spinning, so it is not totally beyond the bounds of possibility that there could exist microscopic black holes with time loops inside. So now, instead of fundamental particles being tiny time machines, we are talking about fundamental particles being tiny time machines inside black holes.

The Idea's Achilles' Heel

Hadley presented his picture of a fundamental particle as a knot of space-time in his Ph.D. thesis, which he defended in 1998. His examiner was gravity expert Chris Isham of Imperial College in London. "At the outset of my defense, he told me he didn't believe a word," says Hadley. "It didn't exactly put me at my ease!"

The grilling from Isham lasted four and a half hours. Midway through, Hadley admits, he was "in despair." Despite everything, however, Hadley was awarded his Ph.D. "A less sympathetic examiner could easily have failed him," says Hyland.

The response from other physicists to the idea has been muted, to say the least. "Hadley has come up with a bold and novel way of reconciling general relativity and quantum theory," says Jonas Mureika of the University of Southern California in Los Angeles. "We can only speculate on whether it is right."

The sentiments are echoed by Hyland. "It's a highly novel idea which at present cannot be proved wrong or right," he says. "I've always been very skeptical but I have to say, it's growing on me."

The idea has yet to grow on Isham. "Hadley's approach is certainly interesting, but it is very speculative," he says. "Most importantly, Hadley has come up with nothing half-resembling a proper theory to give any real substance to his ideas."

Hadley is well aware of this. All he has shown is that general relativity can, in principle, give rise to quantum theory. The phrase "in principle" is important. "I am not yet able to offer an explanation for quantum theory," admits Hadley. "What I have done, however, is show that a gravitational explanation of quantum theory is possible."

The Achilles' heel of Hadley's idea is that he cannot yet show that general relativity permits the existence of a highly localized buckle in space-time, which is his proposed recipe for a fundamental particle. General relativity essentially predicts how a source of energy such as matter warps space-time. Finding the exact shape of the distortion in space-time is referred to as "finding a solution" to Einstein's "field" equations of gravity. Unfortunately, Einstein's equations are horrendously complex, and exact solutions are notoriously difficult to find. It was only in 1962, for instance, almost half a century after Einstein gave the world general relativity, that a solution that describes a spinning black hole was discovered by the New Zealand physicist Roy Kerr.

Hadley is continuing to strengthen his case. Working part-time at Warwick, he has made some progress with a "toy model" of a fundamental particle. This is not enough to convince the skeptics. However, it has some intriguing properties. For instance, his model has a spin and the magnetic properties of an electron. "As yet, I cannot say that it is an electron," says Hadley. "But it looks damned like an electron!"

Hadley may not yet have found a true particle-like solution to general relativity. However, he has done what no one else has done. He has suggested a possible origin for quantum theory, and an origin in old-style classical physics to boot. Einstein never believed quantum theory was fundamental; he thought it was underpinned by something else. One wonders what he would have made of Hadley's ideas. "I think he would have liked some bits of the theory," says Hadley. "But I think he would have hated others—particularly causality violation. However, he was wrong on other predictions of general relativity such as the expanding universe and black holes. He could be wrong on causality violation too."

One can only admire Hadley for having the courage to stick out his neck and risk his career. Whether or not the risk pays off and he becomes the heir of Einstein, only time, or maybe space-time, will tell.

The road Hadley is traveling in his attempt to unify quantum theory and Einstein's theory of gravity is a lonely back road. Meanwhile, others are on an eight-lane highway. They call it "string theory," and it views the fundamental particles of matter as ultratiny pieces of string that vibrate like violin strings. The string theorists are excited because some form of gravity—although not necessarily general relativity—is contained within string theory. The slight complication is that the strings of string theory vibrate in a ten-dimensional world, which means there have to be six extra dimensions that we have somehow managed to overlook.

Extra dimensions have long been the stock in trade of science fiction, the domain of ghosts and of hidden universes. However, the extra dimensions of physics are in an entirely different category. Not only do many physicists believe they exist, some believe they will reveal their presence in experiments within the next few years.

5

Tales from the
Fifth Dimension

Not only are extra dimensions a real possibility,
they could reveal themselves in the next few years

I'm thinking about a fourth spatial dimension, like length, breadth, and thickness. For economy of materials and convenience you couldn't beat it. To say nothing of ground space—you could put an eight-room house on the land now occupied by a one-room house.
> —Robert Heinlein, "And He Built a Crooked House"

Higher dimensional space can be viewed as a background of connective tissue tying together the world's diverse phenomena.
> —Rudy Rucker, *The Fourth Dimension*

It's December 2006, and the Large Hadron Collider near Geneva has just completed its first experimental run. In a dimly lit control room deep underground, a computer display comes alive with the color-coded tracks of two particle "events" plucked from the quadrillions of others seen by one of the LHC's cathedral-size detectors. The effect on the assembled physicists, who have worked themselves to the brink of exhaustion getting the giant experiment up and running, is dramatic. They whoop and cheer and hug each other like long-lost friends.

The events displayed on the split screen each involve the disintegration of a subatomic particle known as a Z-boson into an electron and a positron. At first glance, the two events look indistinguishable. However, a closer inspection shows that the electron and positron in the first event have a total energy of 91 energy units, while their counterparts in the second event add up to 1,091 energy units.* Imagine your surprise if you

---

* The energy units used by particle physicists are in fact gigaelectronvolts.

suddenly came across two mice, one normal and the other weighing as much as an elephant. The physicists who are dancing with jubilation in the LHC control room are a lot more than surprised. The Z-boson that disintegrated in the second event is not just an unusually heavy particle. It's the unmistakable signature of a fifth dimension.

A growing band of physicists is coming to believe in the existence of a fifth dimension—a fourth space dimension in addition to the three of space and one of time with which we are familiar.* And some physicists actually believe that the extra dimension could reveal itself in the next few years when the LHC begins operating at CERN, the European center for particle physics near Geneva.

The idea that there might be a fifth dimension is not new. It arose out of the work of Theodor Kaluza and Oskar Klein in the 1920s. Although the two men were working independently, they were both inspired by Einstein's success in explaining gravity.

What Einstein discovered in 1915 was that there is no such thing as the force of gravity. Imagine a race of ants living on the two-dimensional surface of a taut trampoline. They have no concept whatsoever of the space above and below the trampoline—the third dimension. Now imagine that you or I—mischievous beings from the third dimension— put a cannonball on the trampoline. The ants discover that when they venture near the cannonball their paths are mysteriously bent toward it. Reasonably, they attribute this to a force of attraction that the cannonball exerts on them. From the godlike vantage point of the third dimension, however, it is clear to us that the ants are mistaken. There is no such force. The cannonball has simply made a valleylike depression in the trampoline and this is why all paths curve toward it.

Einstein's genius was to realize that we are in a position remarkably similar to that of the ants on the trampoline. By rights, Earth should be flying through space in a straight line. Instead, its path is constantly bent toward the Sun so that Earth travels in a near-circle. We attribute this Sun-centered motion to a force that the Sun exerts on Earth—the force of gravity. However, if we could only see things from the godlike perspective of the fourth dimension—something that is as impossible for us to do as it is for the ants to see things from the third dimension— we would see that we are mistaken. There is no such force reaching

---

* The three space dimensions correspond to an object's length, breadth, and depth. Alternatively, you can think of them as the directions north-south, east-west, and up-down.

out across empty space. Instead, the Sun creates a valleylike depression in the four-dimensional space-time in its vicinity, and this is why Earth's path curves in a circle around the Sun.*

Einstein's spectacular success in explaining the force of gravity in terms of a higher-dimensional space that we cannot directly perceive encouraged Kaluza and Klein to look to hidden dimensions for an explanation of the other fundamental force of nature—the electromagnetic force, which holds together the atoms in our bodies. They hoped that the electromagnetic force would, like the force of gravity, turn out to be merely a consequence of our limited, antlike perspective on the world.

In a four-dimensional theory like Einstein's, there is room to accommodate only one force—gravity. In order to include the electromagnetic force as well, Kaluza and Klein needed to postulate the existence of one more spatial dimension. Their hope was that, just as the force of gravity could be explained as a manifestation of four-dimensional space-time, the combined forces of gravity and electromagnetism might be explained as a manifestation of five-dimensional space-time.

A Dimension Smaller Than an Atom

People have little trouble accepting that we live in a world with three spatial dimensions moving steadily forward through one dimension of time. However, it is difficult to accept that our universe, rather than possessing four space-time dimensions, is five-dimensional. If there were a fifth dimension, after all, surely we would have noticed it?

Not necessarily.

Think of how a higher dimension might manifest itself to the ants on the trampoline. A mischievous three-dimensional being could reach down, lift an ant off the surface, and put it down somewhere else. To the other ants it would seem like a miracle. An ant would have vanished, and then reappeared at another location. If the ant in question had been confined in an ant prison—and in a two-dimensional world, an ant could be imprisoned merely by having a circle drawn around

---

* Space-time is the name physicists give to the four-dimensional fabric of space and time. The reason for joining the two words into one is that the three spatial dimensions and one temporal dimension are, strictly speaking, not separate. As Einstein pointed out in 1905, space and time are merely different faces of the same coin. It therefore makes sense to talk about living in four-dimensional space-time.

it—the ant warders would have concluded that their charge could walk through walls.

In our world, people simply do not dematerialize in one place and then materialize in another, except on *Star Trek*. Neither do they vanish inexplicably from inside locked prison cells. From this we can conclude that any extra spatial dimensions, if they exist, do not extend very far "above" or "below" our four-dimensional world.* We can be even more specific than this. We can say that any extra spatial dimension certainly cannot extend farther than the width of an atom, which is about a ten-millionth of a millimeter. If it did, air molecules in a sealed room might leak away into the extra dimension, molecule by molecule.

Kaluza and Klein envisioned their extra spatial dimension as "rolled up" into an ultratiny loop much smaller than an atom. According to their picture, every point in normal space is a loop of extradimensional space. The effect of this on the microscopic world is profound. A sub-atomic particle, even when at rest in normal space, can loosely be thought of as traveling ceaselessly around a tiny loop of the fifth dimension like a hamster on a wheel.

The payoff of Kaluza and Klein's extraordinary picture is that it helps explain the electromagnetic force. The source of the force is an electric charge, in much the same way that the source of the gravitational force is mass. In Kaluza and Klein's scheme, an electric charge turns out to be nothing more than the motion of a particle in the extra dimension.

Kaluza and Klein's picture of the two fundamental forces of nature as manifestations of hidden dimensions of space was a beguiling one. Un-fortunately, there was one slight problem. Nature did not have two fundamental forces. It had at least four. In the years after Kaluza and Klein proposed their idea, physicists discovered two "nuclear" forces. The effects of these forces, christened the "strong" and the "weak" force, were apparent only within the central nucleus of the atom, a domain one hundred thousand times smaller than even an atom. This was why they had remained hidden until well into the twentieth century.

A theory of the fundamental forces that explains only half the fundamental forces is hardly a satisfactory theory. Not surprisingly, Kaluza and Klein's ideas were pushed aside. Nevertheless, the physicists' cen-

---

* We use the words north and south, east and west, up and down, past and future, for the directions along the familiar four dimensions. The quotation marks around "above" and "below" are used because we have no comparable words for the directions along the fifth dimension.

tral insight—that nature's forces are simply manifestations of hidden extra dimensions—left a deep impression on physicists. Today's best shot at a unified description of the forces is string theory, and it uses not five dimensions but a total of ten—nine of space and one of time.*

In string theory, four large dimensions are needed to explain the observed properties of gravity, and a further six compact dimensions to explain the electromagnetic, strong, and weak forces. Each point in normal space, rather than being a single loop of extradimensional space, is a loop of six-dimensional space. What such a loop looks like is anyone's guess. The simplest possibility is that it consists of six entirely independent loops, but the likelihood is that the loops are actually intertwined in a far more complex manner, not unlike an intricate knot.

In the Kaluza-Klein picture, the hamster-on-a-wheel motion of a particle in the fifth dimension explained electric charge, the source of the electromagnetic force. In string theory, similar orbital dances in multidimensional space explain the particle properties that give rise to the strong and weak nuclear forces.

If string theory is correct, the obvious question is: where are the extra six spatial dimensions? The string theorists' answer is exactly the same as that of Kaluza and Klein. The extra dimensions are rolled up very small indeed. How small? According to Kaluza and Klein, they are a trillion trillion times smaller than an atom.

Why so fantastically small? It turns out that such a length has a special significance in the world of physics. It is called the "Planck length."

The Planck Length

As Einstein discovered, the warpage of space is gravity. The greater the warpage, the stronger the gravitational force. It is hard to imagine a more warped bit of space than one that is curled back on itself into an ultratiny loop. It follows that the ultracompact extra dimensions of string theory are associated with ultrastrong gravity.

Everyday gravity, however, is far from strong. It is 1 followed by forty-two zeroes weaker than the electromagnetic force that holds together our bodies. The four fundamental forces of nature have wildly different

---

* There are several versions of string theory. Recently, however, physicists have come to recognize that these are merely different faces of an overarching theory. M-theory, as it is called, requires eleven dimensions—ten of space and one of time.

strengths. However, if it were possible to raise the temperature of matter dramatically, greatly boosting the energy available, the strengths of the forces would begin to converge. Physicists believe that, at the ultrahigh energies and ultrahigh temperatures that existed in the first split second of the universe's existence, the forces of nature actually merged, or unified, into a single "superforce."

According to the generally accepted picture of this unification, the three nongravitational forces merge first. Then, at a higher energy, gravity joins the party. The enormously high energy at which gravity becomes so strong that it is at last comparable in strength to the other forces is known as the "Planck energy."

In physics, high energy is synonymous with small distances. This is because subatomic particles have two complementary faces—a particle face and a wave face. Quantum theory, the description of the world of the atom and its constituents, was once widely known as "wave mechanics." This wave character has an important consequence if a particle is confined in a small volume. The wave is scrunched up. Rather than being a spread-out hump of a wave, it becomes a localized, choppy wave. In other words, it becomes more violent, and the energy of the wave increases. This is why the energy associated with the atomic nucleus (nuclear energy) is a million times greater than that associated with atoms (chemical energy). The particles in the nucleus are confined in a box that is a million times smaller than the box of an atom.

Recognizing that small distances are associated with high energies, it is now possible to ask what distance is associated with the Planck energy. The answer is a distance a trillion trillion times smaller than an atom— the extraordinarily small interval physicists call the Planck length.

Here, then, is why the string theorists, and Kaluza and Klein before them, have claimed that the extra spatial dimensions are curled up as small as the Planck length. Curled-up dimensions are associated with ultrastrong gravity. And ultrastrong gravity is found at the Planck energy, which corresponds to the Planck length. The Planck length, claim string theorists, is therefore the natural size for any compact extra spatial dimensions.

Recently, however, some physicists have begun to question this assumption. Strong gravity is undoubtedly associated with the grossly warped space of a curled-up space dimension. However, these physicists say there is no reason to believe it has to be the strongest gravity imaginable—the gravity found at the scale of the Planck length. In 1990, Ignatios Antoniadis of the École Polytechnique in Paris suggested that the extra dimensions might be considerably bigger than the Planck scale. "Apart from the fact that the Planck length is a natural physical scale, there was never any compelling reason why extra dimensions should be of the Planck size," says Keith Dienes of the University of Arizona.

If the extra spatial dimensions are considerably bigger than the Planck length, then an exciting possibility is raised. "The effects of the extra dimensions could be felt by particles at lower energy than the enormous unification energy—perhaps even at energies attainable by the LHC," says Dienes. "It's conceivable that extra dimensions could reveal themselves in the near future."

How exactly would an extra dimension reveal itself? The physicists at the LHC are not going to see particles vanishing into the extra dimensions like rabbits down a hole. The signature of a compact extra dimension is a bit more subtle than that. It was recognized more than seventy years ago by Kaluza and Klein themselves. It has to do with the wave aspect of subatomic particles.

Bigger Than the Planck Length

The wavelike nature of subatomic particles means they are not objects like, say, billiard balls, which are localized at one point in space, but are to some extent spread out like ripples on a pond.* The smeared-out, wavelike nature of subatomic particles has an important consequence if there are compact spatial dimensions. If a particle has sufficient energy—and remember that high energy is synonymous with tiny distances—its "waviness" can actually poke into the extra dimension.

Imagine shouting into a small boxlike room. The sound waves will bounce around in the confined space and create an echo. This is what happens when a particle wave spreads into a small extra spatial dimension. The echo of a sound wave is merely another sound wave. But waves and particles are synonymous. The echo of a particle wave is therefore another subatomic particle, just as solid and as real as the next subatomic particle. "Physicists call them Kaluza-Klein particles and they are the signature of extra space dimensions," says Dienes.

---

* Strictly speaking, we are not talking about a particle itself being smeared out. Rather, the particle has a certain probability of being found in one place, a different probability of being somewhere else, and so on. It is this abstract cloud of probabilities that is spread out.

There is no limit to the number of Kaluza-Klein particles that can exist. Think of the small boxlike room again. Imagine that someone with a higher-pitched voice is shouting. The sound waves will bounce around the room and create an echo, but the echo will be different from the first one. In the same way, when a particle wave with higher energy extends into a compact dimension, it can create a different particle echo. This will correspond to a more massive version of the first Kaluza-Klein particle. Similarly, higher-energy particles can create even more massive Kaluza-Klein particles. "The inevitable consequence of an extra space dimension is therefore the existence of not one or two but an infinite number of Kaluza-Klein particles, each more massive than its predecessor," says Dienes.

How in practice might a Kaluza-Klein particle come into being? First, it is necessary to understand something about energy and mass. Ever since the nineteenth century, physicists have known that one form of energy can be converted into another kind of energy. For instance, electrical energy can be turned into light energy inside a lightbulb; the chemical energy of an explosive can be converted into the energy of motion of a bullet in the barrel of a gun. One of Einstein's greatest discoveries was that mass is also a form of energy—the most compact and concentrated of all forms of energy.* Consequently, not only can mass-energy be turned into other forms of energy, as in a hydrogen bomb, but other forms of energy can be turned into mass-energy. This is the raison d'être of giant particle accelerators such as the one at CERN. They smash together subatomic particles at ultrahigh speeds, converting their enormous energy of motion into the mass-energy of new and exotic particles. Because of the relationship between high energy and small distance, the new particles created are often the building blocks of known particles, enabling physicists to probe the substructure of matter.

If extra spatial dimensions do exist, it follows that a sufficiently violent collision inside a large particle accelerator could provide the necessary energy to conjure Kaluza-Klein particles into being. Every subatomic particle in nature will have an infinite number of Kaluza-Klein cousins. The Kaluza-Klein cousins of the electron, for instance, will be identical in every respect to a standard electron—except for being

---

* The most famous formula in all of science is, arguably, Einstein's $E = mc^2$. This tells us how much energy, E, is stored in a mass, m. The speed of light, c, is a large number. Because it enters the formula squared, the energy in even a small amount of mass is enormous. This is why hydrogen bombs are so destructive despite their small size.

much heavier. There may, for instance, be a Kaluza-Klein cousin of the electron that is two hundred thousand times the mass of a normal electron, another that is four hundred thousand times its mass, yet another that is six hundred thousand times, and so on. And each could be conjured into existence in particle collisions, provided the collisions generated sufficient energy.

The easiest Kaluza-Klein particles to create would be the least massive ones, because they require the least energy to make. So far, however, no particle collision has yielded a single Kaluza-Klein particle. This places a limit on the size of any hidden spatial dimensions. Basically, the bigger the gap in energy between a subatomic particle and its first Kaluza-Klein echo, the smaller the extra dimensions. It's simply that old connection between high energy and tiny distances.

The highest-energy particle collisions that physicists have so far engineered generate a few hundred energy units. That a superheavy version of the lightest particle, the electron, has not turned up in such collisions tells physicists that extra spatial dimensions, if they exist, must be smaller than a billion billionth of a meter in extent. "This is one hundred million times tinier than an atom," says Dienes. "However, compared with the Planck length, it is fantastically huge."

Particle accelerators are getting bigger all the time. Europe's Large Hadron Collider should begin operating in 2006, if all goes according to plan. By slamming together particles with unprecedented violence, the LHC may be able to create a Kaluza-Klein echo. Say, for example, it succeeds in making the lightest Kaluza-Klein echo of the Z-boson particle. The standard Z-boson has a mass of 91 energy units. If an extra spatial dimension exists curled up to a ten billion billionth of a meter, then the separation in energy between the Z-boson and its Kaluza-Klein echoes will be 1,000 energy units. In other words, the Z-boson will have more-massive cousins with energies of 1,091 energy units, 2,091 energy units, 3,091 energy units, and so on. "It's possible that the LHC might be able to make the lightest of these," says Dienes.

A Z-boson is highly unstable and disintegrates in the merest split second into an electron and a positron. Consequently, a superheavy Z-boson, the particle equivalent of an elephant-sized mouse, will disintegrate into a superenergetic electron and a superenergetic positron. This is the scenario described at the outset of this chapter. A superheavy Z-boson, if it is detected at the LHC, will be the unmistakable signature of an extra spatial dimension.

A Submillimeter-size Dimension

An extra dimension that is a mere ten billion billionth of a meter in extent may not seem a lot to get excited about. However, it turns out that unseen dimensions need not necessarily be this small. True, extra dimensions much bigger than this are ruled out by the absence of Kaluza-Klein echoes of familiar particles at the biggest accelerators. However, it turns out that the argument holds only for particles that interact with each other via the three nongravitational forces. "It does not apply to particles that only 'feel' gravity," says Dienes.

What does this mean? First it is necessary to understand how the fundamental forces actually work on a microscopic level. Imagine two tennis players hitting a ball back and forth. Each time one player goes to hit the ball, he or she is knocked back by the power of the opponent's shot. By means of the continual exchange of the tennis ball, therefore, a force is transmitted between the tennis players. In a similar way, a force is transmitted between two microscopic particles by the exchange of "force-carrying particles."

For the electromagnetic force, the force-carrying particles are photons; for the weak force, W and Z bosons; for the strong force, "gluons," of which there are eight different types; and for gravity, a hypothetical particle called a "graviton." Whenever a new particle is created in a particle accelerator, it is the result of the action of one or more of these forces and so involves the exchange of force-carrying particles. The creation of Kaluza-Klein particles is no different. Like normal particles, they come into being because of the exchange of force-carrying particles.

Now consider Kaluza-Klein particles created by the gravitational force. In normal circumstances, gravity is fantastically weak compared to the other forces of nature. In the language of force-carrying particles, this means that particles in accelerators hardly ever exchange gravitons. Gravity therefore plays essentially no part in the processes that go on there. Consequently, even if Kaluza-Klein particles were created by the action of gravity, their effects would be minuscule and undetectable. The failure to see the effects of Kaluza-Klein particles made by gravity cannot therefore be used to conclude anything about the size of extra spatial dimensions. "The upshot is that there could be extra dimensions experienced by gravity, and gravity alone, and they could be considerably bigger than a billion billionth of a meter in extent," says Dienes.

How, then, might we discover such a gravity-only dimension? The direct detection of Kaluza-Klein echoes is only one way of spotting a hidden spatial dimension. Another way is to look for indirect effects. In the case of gravity, these might not be too difficult to spot.

Extra spatial dimensions make possible Kaluza-Klein echoes of all subatomic particles. This means there can be Kaluza-Klein echoes of the force-carrying particles too. If there is enough energy around, we can therefore expect to find Kaluza-Klein echoes of, say, the photon of the electromagnetic force or the gluons of the strong nuclear force. These will be bound to have an effect. After all, if a particular force has more force carriers than usual, the force is sure to be modified.

That Kaluza-Klein echoes of the force carriers would change the strengths of the four fundamental forces was one of the chief reasons why, until recently, physicists were unwilling to consider the possibility of extra dimensions bigger than the Planck length. They believed that at high energies, the extra force carriers would modify the forces in such a way that they would never unify into a single superforce. It was the discovery in 1998 by, among others, CERN physicists Keith Dienes, Emilian Dudas, and Tony Gherghetta that this may not happen—that unification might actually occur at much lower energies—that stimulated the current interest in possible extra dimensions bigger than the Planck length.

It is the modification of gravity by extra Kaluza-Klein gravitons that could reveal an extra gravity-only dimension. Normally, the force of gravity between two massive bodies gets weaker with the square of the distance between them. When two massive bodies are twice as far apart, for instance, the force is four times weaker; when they are three times as far apart, it is nine times weaker.

Extra gravitons would be expected to modify this "inverse-square law." However, we see no such modifications on the large scale. Gravity is perfectly well behaved in the solar system and even on the scale of physics laboratories. However, nobody has tested the inverse-square law with bodies that are less than about 0.2 millimeter apart. So although we can say with certainty that any extra dimensions experienced by the three nongravitational forces must be smaller than a billion billionth of a meter in extent, we can say only that any extra dimensions experienced by gravity must be less than 0.2 millimeter.

The incredible possibility that there might exist submillimeter-size extra dimensions that we have overlooked was pointed out in 1998 by

Nima Arkani-Hamed and Savas Dimopoulos at Stanford University
and Gia Dvali of the International Centre for Theoretical Physics in
Trieste, Italy.

If such a gravity-only dimension exists, Kaluza-Klein echoes of gravitons could make gravity millions of times stronger than the inverse-
square law predicts on scales of less than 0.1 millimeter. Similar ideas
could even make gravity repulsive. Incredibly, this does not conflict
with any current observations we have made of the universe.

Experiments are currently under way at Stanford University, the
University of Colorado, and the University of Washington in Seattle to
check the inverse-square law of gravity on scales much less than 1 millimeter. The experiments are difficult to perform. However, they hold
out the amazing possibility of finding an extra dimension with a relatively simple tabletop experiment.

Consequences of Big Extra Dimensions

If extra dimensions much bigger than the Planck length do indeed exist, what would be the consequences? For one thing, particle physicists
would have a whole new zoo of particles to study, which would be
bound to make them happy. Also, as Dienes, Dudas, and Gherghetta
have shown, the extra force carriers could bring about the unification
of the forces at an energy much lower than expected. This is important
because the details of unification are hazy. For instance, physicists
think that unification is brought about by a whole new array of superheavy force-carrying particles called X- and Y-bosons. These particles
may be involved in processes that favor matter over antimatter and so
could at last explain why we live in a matter-dominated universe.

All subatomic particles have associated with them antiparticles with
opposite properties such as electrical charge. For instance, the negatively charged electron is twinned with a positively charged antiparticle
known as a positron. When a particle meets its antiparticle, the two
vanish in a burst of radiant energy. Because particles and their antiparticles are always created together, one of the outstanding mysteries of
physics is why we appear to live in a universe of matter and not one
with a fifty-fifty mix of matter and antimatter.

The energy at which unification happens is so fantastically great that
to observe the process physicists would need a particle accelerator a

hundred trillion times more powerful than any in existence. However, if there are extra dimensions substantially larger than the Planck length, the unification energy could be lowered dramatically so that unification could be observed in the near future. "In fact, if the extra dimensions are rolled up ten times smaller than any scale we have yet probed—a ten billion billionth of a meter—the unification energy could be observed at the LHC," says Dienes.

The consequences of a big extra dimension noticed only by gravity are even more amazing. Conceivably, the energy at which gravity merges with the three nongravitational forces could be brought down within reach of accelerators. That would mean we could examine gravity when it is comparable in strength to the other forces—in other words, when it could drastically warp the fabric of space and time.

If space is warped enough, it results in a black hole, a region of space where gravity is so powerful that even light is trapped. If time is warped enough, the result might be a looping back of time on itself—a time machine. By sufficiently warping space-time, it might be possible to create a new universe or even a wormhole, a tunnel connecting widely separated regions of space and providing a handy shortcut for interstellar travelers.

Black holes, time machines, new universes, and wormholes. If gravity merges with the other fundamental forces at reachable energies, we may be able to create all of these wonders by manipulating the fabric of space-time inside particle accelerators.

But if these possibilities are not mind-blowing enough, there is one other surprising possibility. It is so surprising that it almost defies belief. There may exist an extra dimension experienced by gravity that isn't a mere 0.2 millimeter in extent but that is infinite, just like the familiar space dimensions.

An Infinite Extra Dimension

But wait a minute—isn't there an experimental limit of 0.2 millimeters? Yes, there is. Incredibly, however, there may be a way to have an infinite extra dimension that gravity nevertheless penetrates to a depth of less than 0.2 millimeters. It all has to do with string theory.

In string theory, the forces of nature are manifestations of ten-dimensional space-time. The theory allows for the possibility of lower-dimensional islands, or "branes," floating within this higher-dimensional space. Conceivably, our universe is a four-dimensional brane suspended within a ten-dimensional space-time.

Until recently, nobody thought much about the dimensions that extended beyond the four-dimensional brane. The gravity of objects within the brane such as galaxies and stars and planets would inevitably spread beyond the brane. This would dilute gravity on the brane so that it fell away faster than if it followed the inverse-square law we observe.

However, according to Lisa Randall of Princeton and Raman Sundrum of Stanford, this need not happen. We could live on a four-dimensional brane with at least one of the extra six spatial dimensions infinite in extent. What would stop gravity spreading out into the extra spatial dimension, according to Randall and Sundrum, is the gravity of the brane itself. This is enormously strong and warps space so drastically that the gravity of objects within the brane such as stars cannot extend more than 0.2 millimeters from the brane.

Gravity can therefore extend an infinite way into the extra dimension, yet still be scrunched up within a millimeter of the brane. "Incredibly, an infinitely big space dimension could have gone completely unnoticed until now," says Sundrum.

One of the things all these ideas about extra dimensions teach us is that there is more to the universe than we imagined in our wildest dreams. And a possible fifth, sixth, or nth dimension is only the beginning. Astronomers have discovered that most of the matter in the universe is not tied up in familiar stars and galaxies but is in some invisible form that reveals its presence only by tugging on the visible stuff with gravity. Nobody really knows what the invisible material is. However, there have been hundreds of suggestions. Among the weirdest of all is the proposal of an astronomer based in Scotland. His name is Mike Hawkins and, if he is right, most of the mass of the universe is in the form of black holes the size of your refrigerator.

Part Two
The Nature of the Universe

6

The Holes in the Sky

*Is most of the mass of the universe
in the form of refrigerator-sized black holes?*

> The black holes of nature are the most perfect macroscopic objects there are in the universe: the only elements in their construction are our concepts of space and time.
>
> — Subrahmanyan Chandrasekhar

You would never guess it was there. A tiny blob of impossible blackness floating against the jeweled backdrop of stars. It has hung there since the beginning of time, a knot of tortured space, no bigger than a small refrigerator but as heavy as a full-grown planet. What in the world is it? Simple. A miniature black hole spawned in the first split second of the universe's existence. If an astronomer in Scotland is right, black holes like this may account for an astonishing 99 percent of the mass of the entire universe.

Why would anyone believe such an extraordinary thing? Because it could explain something peculiar about "quasars." Quasars are the superbright cores of newborn galaxies. Typically, they shine brighter than a hundred normal galaxies such as our Milky Way, making them easily visible even at the edge of the universe.*

In 1975, when Mike Hawkins began a long-term project at the U.K. Schmidt telescope at Siding Spring Observatory in Australia, quasars were the last thing on his mind. What interested the astronomer from

---

* Actually, there are relatively few quasars around today. Their heyday was when the universe was young. Astronomers nevertheless use the present tense in discussing quasars — and everything else in the night sky — because we are seeing them "today." Quasars are long dead but their light, after its multi-billion-year journey across the universe, continues to arrive on Earth.

the Royal Observatory in Edinburgh was variable stars. Variable stars are suns that brighten and fade, usually over a period of a month or so, because of some instability deep in their interiors. Hawkins's plan was to find new variable stars by using the Schmidt telescope to monitor continuously a particular portion of the sky over a period of many years.

A Schmidt telescope is particularly suited to such a search because it has an unusually large field of view compared to an ordinary telescope, which zooms in on a tiny bit of the sky. The Schmidt telescope magnifies a region of the sky 4.5 degrees on a side, which is equivalent to nine Moon-widths.

In one respect, Hawkins's plan worked perfectly. By 1980, he had found lots of celestial objects that fluctuated in brightness. The surprise, however, was that most of them were not stars. They were quasars.

That quasars vary the light they give out is well known. It was precisely this property that, soon after the discovery of quasars in 1963, alerted astronomers to the extraordinary nature of these prodigiously luminous objects. Despite pumping out as much energy as a hundred normal galaxies, they are very, very small—typically smaller than our solar system.

The connection between the rapid changes in brightness of an object and its size is because of the cosmic speed limit imposed by light. Say, in a corner of a quasar, there is a sudden release of energy—an explosion, perhaps—that causes the region to heat up and brighten dramatically. Inevitably, the energy will spread to neighboring regions, causing them to heat up and brighten too. However, the energy of the explosion cannot spread faster than the speed of light.* The speed of light, according to Einstein, is the ultimate cosmic speed limit. Consequently, if a quasar takes a few weeks to climb in brightness, we can say with confidence that it cannot be any larger than the distance light travels in a few weeks, which is about a hundred million million kilometers.

In addition to visible light, quasars commonly give out X rays, which are produced by matter heated to millions of degrees. Often, the X-ray glow of quasars brightens and dims in the space of only a few

---

* The speed of light is three hundred thousand kilometers a second—a million times faster than sound. At this speed, human distances—for instance, the distance from a tree to your eye—are spanned in an instant. To all intents and purposes, we therefore see the world around us "at this moment." On the other hand, astronomical distances are huge and light's progress across them is positively snail-like. Consequently, we see even the nearest stars as they were years ago and the most distant quasars as they were many billions of years ago.

hours, implying that the region producing the X rays is smaller even than the distance light can travel in a few hours. This is about the size of the solar system.

What possible "engine" can generate as much energy as a hundred normal galaxies—a total of ten million million suns—from so small a volume of space? Ever since their serendipitous discovery by the Dutch astronomer Maarten Schmidt, this has been the central puzzle of quasars. Nuclear energy—the energy that keeps the sun hot and provides us with sunlight—falls far short of the requirements. Only one known process can come up with the goods: the infall of matter into a black hole.

A black hole is a region of space where gravity is so strong that nothing, not even light, can escape. A region of space is just about all a black hole is. The space is so grossly warped that it can best be compared with a bottomless well down which light and matter pour. Because a black hole is not anything solid, theorists talk merely of the "horizon" surrounding a black hole. This imaginary membrane marks the point of no return for infalling light and matter. Once inside, nothing can climb out again.

In the standard model of a quasar, a black hole lurks in the heart of a young galaxy. But we are not talking about an ordinary black hole—that is, one as massive as a few suns. We are talking about a "supermassive" black hole, a monster weighing anything up to ten billion times the mass of the Sun.* Interstellar gas and ripped-up stars spiral down into the hole like water going down a drain. Friction within this inswirling "accretion disc" of matter heats it up to millions of degrees so that it glows blindingly bright. This is the ultimate source of the extraordinary light output of a quasar.

The picture of the accretion disc allows us to be a little more precise about the reason some quasars can fluctuate in brightness in as little as a few hours. It takes a few hours for the effect of any disturbance, propagating at the speed of light, to travel from one end of the accretion disc to the other. Here, at last, we come to the puzzle presented by

---

* The strong suspicion is that all galaxies, including our own Milky Way, passed through a troublesome quasar phase in their adolescence. Certainly, there is evidence of large but quiescent black holes hiding in the cores of several nearby galaxies. In fact, our own galaxy appears to have one—a several-million-solar-mass extinct quasar known as Sagittarius A. Why do quasars stop being quasars? A good bet is that they run out of matter to fuel their enormous appetites.

Hawkins's observation program. The quasars in his sample turned out not to be brightening and fading in the space of a few hours or a few days, or even a few weeks. They were varying their light output over periods of five to ten years.

The dimming and brightening was smooth and typically amounted to a 30 to 100 percent change in brightness. By contrast, the variability of quasars on a timescale of a few hours involved them changing their brightness by only a few percent.

Nobody has a good idea of what could cause the kind of long-term changes in quasars found by Hawkins. It could be that there are slow changes in the rate at which the surrounding galaxy feeds matter into the hungry quasar at its heart. But then that leaves the mystery of what causes those changes. Most astronomers, however, are not worried unduly by their ignorance. Quasars, they point out, are deeply mysterious objects, and our knowledge of them is still rudimentary. In the fullness of time an explanation for the long-term variability is bound to turn up inside quasars.

Hawkins disagrees. He believes that an explanation will not be found by looking inside quasars or inside the galaxies that feed quasars. Controversially, he thinks that the long-term variability has nothing at all to do with quasars.

But if it is not caused by something on the inside of quasars, then surely it must be caused by something on the outside. Precisely. Quasars are varying their light output over the long term, Hawkins maintains, because there are bodies floating in the space between us and the quasars and they are sailing across our line of sight, as does an airliner passing in front of the sun. The difference is that, whereas an airliner temporarily dims the Sun, the intervening bodies actually boost the light of a quasar.

Microlensing

This effect is known as "gravitational lensing." It occurs because the gravity of a massive body actually bends the path of any light that passes close to it. The reason gravity does this, as Einstein discovered, is that gravity is a warpage of space. A massive body such as the Sun creates a sort of dimple in the space around it, and the path of any light passing near is bent merely because the light must negotiate the dimple.* If the gravitational lensing is caused by a body between us and a distant object such as a quasar, the overall effect is to focus and magnify the light of the distant object in much the same way as does a converging glass lens.

But what made Hawkins think that the long-term variability was caused by something in space passing in front of the quasars? Several things. One was simply the difficulty of explaining slow changes in brightness over many years when light takes only hours to cross the accretion disc of a typical quasar. A second was that the quasars did not change their overall color as they varied their brightness. Usually, when an astronomical body brightens or fades, it changes its color as well. This is because a change in brightness is normally caused by a change in temperature. Think of how a piece of metal changes from white-hot to dull cherry red as it cools down.

Hawkins's monitoring program involved looking through different-colored "filters"—red, blue, and ultraviolet (a type of invisible light responsible for sunburn). A red filter, not unlike a piece of red cellophane, allows through only the red light of an object, a blue filter only the blue light, and so on. What Hawkins noticed was that the red light from each quasar went up and down in perfect step with the blue light and the ultraviolet light. Because the mix of individual colors determines a body's overall color, this meant that the overall color of the quasars did not change. "A lack of any color change turns out to be a characteristic feature of gravitational lensing by an intervening massive object," says Hawkins. "According to Einstein, the individual colors of light are bent, or magnified, by precisely the same amount."

Another characteristic of gravitational lensing is that the variation in light is "symmetric" in time as the lensing body passes in front of the distant object. What this means is that the light climbs in brightness in exactly the same way that it falls. This is only the case, however, if there is a single massive body between us and a distant quasar. If, as Hawkins believes, all quasars are being "lensed," it is unlikely that each has pre-

---

* In 1919, this effect was used to confirm Einstein's general theory of relativity. British astronomers took advantage of a total eclipse of the Sun, in which the Moon passes in front of the Sun and blots out its glare, to observe stars close to the solar disc. The light from the stars, on its journey to Earth, had to negotiate the gravitational dimple around the Sun and so ought to have had its path bent from a straight line. Sure enough, the astronomers found the light deflection predicted by Einstein.

cisely one lensing body passing in front of it. Some could easily have several. Unfortunately, lensing by multiple bodies is confusing. "You do not get the simple symmetric variation in light which is the 'finger-print' of gravitational lensing," says Hawkins.

Despite this, it is possible to look at a large sample of quasars and see whether, on average, they fade in the same way that they brighten. "It's not as convincing a test as seeing the effect in a single quasar," says Hawkins. "Nevertheless, the result is very suggestive of gravita-tional lensing."

What Are the Lensing Bodies?

The case for gravitational lensing bodies is not yet watertight. However, Hawkins believes that the weight of the evidence points to the exis-tence of bodies between Earth and the quasars that temporarily boost the light of quasars as they pass in front of them. What are these myste-rious bodies? A clue is found in the time a quasar takes to fluctuate up and down in brightness. This turns out to be linked to both the mass of the lensing body and precisely where the body is located on the line between the quasar and us. "Of course, we cannot know for sure where they are but a safe estimate is halfway," says Hawkins. "Making this as-sumption, it is possible to come up with an estimate of the mass of the lenses. They are roughly as heavy as the planet Jupiter."

What are these mysterious Jupiter-massed bodies? The answer, says Hawkins, came from an unexpected quarter. In the 1980s, David Schramm and Matt Crawford of the University of Chicago had been investigating the birth of black holes.* A black hole usually forms when a massive star runs out of nuclear fuel at its core. A star is a ball of hot gas in which the outward force of the hot gas opposes the inward force of gravity trying to crush it. If it runs out of fuel with which to generate heat, the core shrinks catastrophically under the pressure of gravity. Quickly, it becomes so compressed and its gravity so strong that not even light can escape. This is the standard way of making a black hole—by suction from inside. But there is another possible way: by squeezing from outside.

---

* David Schramm was one of the world's foremost theoretical astrophysicists. Unfortu-nately, he died in an airplane crash in 1998.

Such a process almost certainly cannot happen in the universe to-day. However, Schramm and Crawford realized that it might have occurred in the past—in the first few moments after the birth of the universe. As the fireball of the Big Bang expanded and cooled, it passed through a phase so violently turbulent that it could easily have spawned black holes. This was the "quark-hadron phase transition."

Before that time, a mere millionth of a second after the moment of creation, the universe was so hot that quarks were flying about too frantically ever to stick together. After that time, they had slowed down enough to form clumps. The protons and neutrons that make up the atoms of our bodies are actually little "bags" of quarks—one combination of three quarks makes up the proton and another combination the neutron. The quark-hadron phase transition is therefore the moment in the universe's history when quarks condensed into neutrons and protons.

This event was accompanied by a huge release of energy. A more familiar example of a phase transition is the condensation of steam into water droplets. To convince yourself that this releases energy you need only put your hand in the steam from a kettle and see how it scalds. Schramm and Crawford realized that so much energy was released during the quark-hadron phase transition that it violently stirred up the universe. Some regions could easily have been squeezed so brutally that they became dense enough to form black holes.

Schramm and Crawford proceeded to calculate how big the black holes would have been. Their answer? Roughly the mass of Jupiter. "They were just the right mass to be the mysterious bodies which were lensing all quasars," says Hawkins.

A black hole of this mass would be only a meter or so across—about the size of a small refrigerator. In 1975, Hawkins had set out with the modest aim of discovering new variable stars. Now, bizarrely, he seemed to have discovered that the universe was chock-a-block full of refrigerator-sized black holes formed in the first split second of creation.

The question that immediately occurred to Hawkins was: how much of the universe's mass was tied up in these objects? Oddly enough, the relevant calculation had already been done. In 1973, James Gunn and William Press of the California Institute of Technology in Pasadena were thinking about the phenomenon of gravitational lensing and wondering about the effect it might have on the images of distant quasars. The lensing bodies they had in mind were galaxies between us and the quasars, not Jupiter-massed black holes. However, the conclusion they arrived at turned out to depend not on the details of the individual lensing bodies, but only on their total mass. If the bodies were large, each would magnify the light of quasars over a large area of the sky. If they were small, each would magnify a smaller area, but this would be exactly compensated for by a larger population of objects. "Gunn and Press came to a particularly simple conclusion," says Hawkins. "Every quasar would be lensed if the total mass of the lensing bodies was equal to the universe's 'critical mass.'"

The concept of "critical mass" is of central importance to our understanding of the long-term fate of the universe. Currently, the universe is expanding, its constituent galaxies flying apart like pieces of cosmic shrapnel in the aftermath of the Big Bang. However, the primordial explosion is not the only thing influencing the motion of galaxies. The galaxies themselves are pulling on each other with the force of gravity, and the combined effect of this is gradually to brake the expansion of the universe. The total braking force depends on the total mass of the galaxies and other material in the universe. And this is where the notion of critical mass comes in. If the total mass of the universe is less than the critical mass, gravity cannot halt the expansion and it will continue forever; if the total mass is more than the critical mass, the universe will one day recollapse down to a Big Crunch.

The dividing line between these two possibilities is a universe at exactly critical mass. In this case, the universe's expansion will run out of steam painfully slowly and come to a halt only after an infinite amount of time has elapsed. For various technical reasons, astronomers believe we live in such a universe.* This presents them with a major headache, because the galaxies they can see with their telescopes add up to only about 1 percent of the critical mass. To hide their embarrassment, they postulate that the rest of the universe's material is in some mysterious, invisible form called "dark matter."

Evidence for dark matter comes from other places too. For instance, the stars in our Milky Way are whirling around the center of the galaxy

---

* One reason is that a universe that starts out with less than the critical mass becomes ever more rarefied as it expands, and this means it gets farther and farther away from the critical mass. Similarly, one that starts out with more than the critical mass contracts and becomes ever more dense. This means it too gets farther and farther away from the critical mass. It's rather like a pencil balanced on its point. If it starts off even a slight amount away from the vertical, it will get farther and farther away from the vertical—that is, fall over. The only way it can stay balanced is if it starts off perfectly vertical. Similarly, the only way the universe can stay close to the critical mass is if it starts off at the critical mass.

so fast that they ought to be flung off into intergalactic space. They stay in orbit because they are in the grip of the gravity of a large amount of invisible, dark matter. There must be at least ten times more mass tied up in dark matter than there is in visible material such as stars. Those keeping track of the figures will realize that this still leaves quite a bit more missing mass in order to get to the critical density. The current suspicion is this is accounted for by the energy of empty space.

According to Hawkins, however, the mystery of the dark matter is a mystery no more. If the universe is filled with black holes with the mass of Jupiter, and they lens every single quasar, Gunn and Press's result implies that the total mass of such holes must be close to the critical mass.

This is an extraordinary conclusion. If Hawkins is correct, refrigerator-sized black holes that formed during the Big Bang are the dominant component of the cosmos. They account for 99 percent of the mass of the cosmos and, through their combined gravity, control the fate of the universe.

Should we be worried? After all, black holes suck in matter remorselessly. Knowing their total mass and the size of the universe, it is a straightforward matter to work out how close the nearest one is to Earth. "The nearest mini black hole should be thirty light-years from Earth, which is almost ten times farther away than the nearest star, Alpha Centauri," he says. "We ought to be safe."

In coming up with the thirty-light-year figure, however, Hawkins assumed that the black holes are spread uniformly throughout space. But this may not be the case. They may instead be clumped around galaxies like our own Milky Way. If this is the case, the nearest could be a lot nearer. Let's hope it's not too near, however. After all, if a refrigerator-sized black hole were to pass near Earth, it would pull catastrophic tides in the oceans. If it were to lodge inside Earth, it could devour the planet from within—the ultimate cosmic vacuum cleaner.

Most astronomers believe that most of the matter of the universe is in some invisible form. If you balk at the idea of refrigerator-sized black holes, however, you may have to leave the room at this point. For two physicists are suggesting that the invisible stuff could be in a form that is far more exotic. If Robert Foot and Sergei Gninenko are right, there could be an entire invisible universe occupying the same space as the visible universe. They are suggesting the existence of invisible galaxies, invisible stars, and invisible planets—even invisible ETs.

7

Looking-Glass Universe

*Our universe could contain invisible galaxies, stars,
and planets—even invisible ETs*

"Paul?" asked Elizabeth, looking perplexed. "How could another
planet occupy the same space as the Earth?"
—John Cramer, *Twistor*

Mirror, mirror on the wall, who is the fairest of them all?
—The Wicked Witch, *Snow White*

In an invisible galaxy far, far away, an invisible swarm of stars wheels
about an invisible galactic nucleus. Many of the invisible stars are
circled by invisible planets, and on some of the planets life goes about
its business—you've guessed it—in total invisibility.

According to some physicists, our universe may contain an entirely different kind of matter besides the familiar stuff we know and love. It goes
by the name of "mirror matter" because, in a loose sense, it is like ordinary
matter reflected in a mirror. Except that it is invisible—so completely invisible that even if our universe were chock-a-block full of mirror galaxies,
mirror stars, and mirror planets, we would never have noticed.

But why believe such a bizarre thing? Because it would repair a
damaged symmetry of the universe. Symmetry concerns the things that
do not change when an object is transformed in some way. For instance, a face that looks the same when it is reflected in a mirror is said
to show "mirror symmetry." Similarly, a starfish that looks the same
when it is rotated through a fifth of a complete turn is said to exhibit
"rotational symmetry."

Symmetry has been a guiding light in the search for the fundamental laws that underpin the cosmos. Nature, for reasons best known to it-

self, has chosen laws that exhibit the maximum possible degree of symmetry. For instance, the laws of physics are the same today as they were yesterday—in other words, they are symmetric with respect to transformations in time. The laws are also the same in London as they are in New York—that is, they are symmetric with respect to transformations in space. And in addition, the laws respect a myriad of other, more abstract symmetries.*

But although nature demonstrates a fanatical devotion to symmetry, its symmetry is not quite perfect. In particular, there is one situation in which it is curiously flawed: when the laws of physics are reflected in a mirror.

Consider microscopic particles such as electrons and neutrinos, quarks and photons. The laws of physics permit these particles to interact in a bewildering number of ways. For instance, an electron can spit out a photon of light that can then be absorbed by another electron; a quark can emit a particle known as a vector boson and turn into another kind of quark. Now imagine that these processes are reflected in an imaginary mirror. If the laws that govern how particles interact are mirror-symmetric, then in every instance the mirror-reflected process will turn out to be one that can occur in nature.

So, do all mirror-reflected processes happen? The remarkable answer, discovered by the Chinese-American physicists Tsung Dao Lee and Chen Ning Yang in 1956, is no. Sometimes, the mirror-reflected process is one that is never, ever seen in nature. Take a particle known as a neutrino. When it is born, it always corkscrews in a left-handed manner about its direction of travel. No experiment has ever caught a neutrino corkscrewing in a right-handed manner.

This is surprising. Think of a soccer team whose players pass the ball back and forth until, finally, one of them scores a goal. Now imagine the same sequence of passes reflected in a hypothetical mirror. For instance, the mirror might be along one side of the field so that all the passes to the left become passes to the right, and vice versa. Could the sequence of passes seen in the mirror actually occur in the real world? The answer is yes. As long as the ball remains on the field of play, the

---

* Every symmetry of the laws of physics is associated with a quantity that never changes or, to use the technical term, is "conserved." For instance, that the laws of physics are the same at different times implies that a quantity known as "energy" can never be created or destroyed. Similarly, that the laws are the same in different places implies that a quantity known as "momentum" must always stay the same.

laws of soccer place no restriction whatsoever on the direction in which it is passed.

Imagine how bizarre it would be if the laws did impose a restriction, and that in some situations a player could kick the ball to the left but not to the right; that echoes the peculiar state of affairs that exists in the microscopic world.

Mirror Particles

Lee and Yang were deeply troubled by their discovery. In experiments done before 1956, nature had displayed an absolute passion for symmetry. Why, in the case of mirror reflection, was that symmetry flawed? Lee and Yang made a daring suggestion. Perhaps nature's symmetry was not flawed. Perhaps it only looked that way.

There was only one way to restore nature's left-right symmetry. According to Lee and Yang, there must exist a "mirror" or "shadow" world, complete with mirror-reflected particles. Both types of neutrino would then exist—left-handed ones in our world and right-handed ones in the mirror world. The only reason that nature's left-right symmetry appeared flawed, claimed Lee and Yang, was that we were not able to see the mirror world.

Mirror particles would be identical in most respects to their ordinary counterparts. For instance, a mirror electron would have the same mass as an electron, a mirror up-quark the same mass as an up-quark. The principal difference would be that they would interact with each other in ways that would be the mirror image of the ways in which their ordinary counterparts interacted. In most cases, this back-to-front interaction would be the only noticeable difference. Only in rare instances—for example, in the case of mirror neutrinos, which would have to corkscrew the opposite way in order to interact in a mirror-image fashion—would the difference be marked.

Mirror particles would differ from ordinary particles in another rather important respect: they would be invisible to detection. Why? Well, ordinary particles interact with each other by exchanging particles. This is the microscopic basis of all forces. For instance, the electromagnetic force arises from the continual exchange of photons, "force-carrying" particles that are batted back and forth rather like tennis balls between two tennis players. Nature's strong nuclear force arises from the exchange of gluons, and so forth.

For mirror particles to have failed to show up in any experiments to date, they must entirely ignore ordinary particles. In other words, they must interact with none of the known force-carrying particles. The failure to interact with the carriers of the electromagnetic force—photons of light—would explain why we do not see mirror matter. A body has to interact with light in order to reflect light into our eyes so that we can see it.

However, mirror particles, although they ignore the known forces, must nevertheless interact with one another. And this requires the existence of a whole new set of forces—mirror forces. For instance, there must be a mirror electromagnetic force carried by mirror photons, a mirror strong nuclear force carried by mirror gluons, and so on. And ordinary particles ignore mirror forces as conscientiously as mirror particles shun standard forces.

The price of fixing nature's flawed left-right symmetry is therefore a whole zoo of mirror particles that interact via mirror forces carried by mirror force carriers. Everything in our world would be duplicated in the mirror world. For instance, mirror quarks would assemble themselves into mirror protons and neutrons, which in turn would make mirror atoms and molecules. And there is no reason to believe there could not be mirror galaxies, mirror stars, mirror planets—even mirror living organisms.

The Evidence for Mirror Matter

Back in 1956, when Lee and Yang suggested that nature's imperfect symmetry could be restored by a mirror world, few took the idea seriously. More than four decades later, however, things have changed a little. Two respected physicists are actually claiming that there is laboratory evidence for the existence of a mirror world.

Robert Foot of the University of Melbourne and Sergei Gninenko of CERN point to a puzzling observation made in 1990. In an experiment carried out at the University of Michigan at Ann Arbor, physicists measured the lifespan of something called "ortho-positronium."

Positronium is the simplest atomlike system. It consists of an electron and a positron orbiting each other in the grip of their mutual electric force.* Physicists make it by firing a low-energy beam of positrons into matter. Some of them slow down enough to steal electrons away from ordinary atoms and form positronium.

Not all positronium, however, is the same. Electrons and positrons have a property known as "spin," which, crudely speaking, is like the spin of a top. When an electron and positron come together to make positronium, they have two options open to them. They can either both spin the same way, which makes ortho-positronium, or they can spin in opposite ways, making "para-positronium."

Ortho-positronium is a particularly simple entity whose properties can be precisely predicted by the theory of Quantum Electrodynamics. QED is the theory of the electromagnetic force, which in essence means it is the theory of how light interacts with matter. The theory explains almost everything about the everyday world, from why the ground beneath our feet is solid to how a laser works, from the chemistry of metabolism to the operation of computers.

QED predicts that, after living for 142 billionths of a second, an average ortho-positronium should self-destruct, or "decay," into three photons. This may not seem very long. However, the Michigan experimenters were able to time the interval with ultrasensitive instruments. And they discovered that the decay of ortho-positronium occurred 0.1 percent faster than it should have. This might appear a tiny discrepancy. However, QED is renowned for its phenomenal accuracy in predicting the outcome of experiments. The difference was therefore significant.

Nobody worried too much at the time. They assumed that if theorists improved their calculations, taking into account what are dubbed "higher order radiative corrections," the lifetime predicted by QED would be found to match the lifetime observed in the Michigan experiment. The improved calculations were carried out in March 2000 by Gregory Adkins of Franklin and Marshall College, R. Fell of Brandeis University, and Jonathan Sapirstein of Notre Dame University. To everyone's surprise, the discrepancy remained.

---

* The positron is the "antimatter" twin of the electron. All ordinary particles have antimatter counterparts. For instance, the proton is twinned with the antiproton, the neutrino with the antineutrino. One of the greatest mysteries of physics is why we appear to live in a universe made exclusively of matter when the laws of physics seem to predict a fifty-fifty mix of matter and antimatter. Antiparticles such as the positron are frequently observed in experiments at particle accelerators. Because mirror particles mirror ordinary particles, they have their own antiparticles—the mirror positron, mirror antiproton, and so on.

Enter Gninenko and Foot. In a paper published in the 11 May 2000 issue of the journal *Physics Letters B*, the two physicists pointed out that there was a way to explain the puzzle of the ortho-positronium lifetime. It required the existence of a mirror universe.

A Probe of the Mirror World

How could the lifetime of ortho-positronium possibly be affected by the presence of a mirror universe? Mirror matter, after all, shuns ordinary matter. However, in 1985, Bob Holdom, a physicist at the University of Toronto, pointed out that the two types of matter might interact with each other through a hitherto unknown force.

Holdom's reasoning was as follows. One of the guiding principles of physics is that, underneath all the complexity we see around us, nature is actually quite simple. This is an act of faith — we do not know why the universe should be built this way — but, over the years, this faith has been rewarded with an unprecedented understanding of the world and an unprecedented degree of control over it. The belief that, deep down, things are simple prompts today's physicists to argue that the four fundamental forces of nature are merely facets of a single superforce that ties together all particles. If mirror matter exists, therefore, nothing could be more natural than to postulate a force that ties together ordinary matter and mirror matter into a single, unified framework.

Holdom's hypothetical force would have to be very weak or else we would already have spotted its effects. Recall that the forces between particles arise from the exchange of force-carrying particles. There must therefore exist a force carrier of the new force. Let's call it an "H particle" for simplicity. The exchange force arises when an ordinary particle spits out an H to become a mirror particle and a mirror particle gains an H to become an ordinary particle. The process also happens in reverse, with the H particle batted back and forth like a tennis ball.

Now it turns out that the force between matter and mirror matter has an important characteristic that follows from something called the "conservation of electric charge." This is the idea that electric charge, like energy, can never be created or destroyed. The amount at the beginning of any process is always the same as the amount at the end. Take an electrically charged particle such as an electron. If an electron spat out an H and became a mirror electron, it would end up with no charge, because mirror electrons do not have ordinary charge—they have mirror charge. But destroying an electric charge is forbidden. Similarly, if a mirror electron gained an H and became an ordinary electron, an electric charge would have been created, which is also forbidden.

So there can be no interaction that changes an ordinary particle with a charge into a mirror particle. Such an interaction can arise only if a particle has zero electric charge and a mirror particle has zero mirror charge (for everything that has been said so far applies to mirror particles, which are as incapable of losing or gaining mirror charge as ordinary particles are of losing or gaining ordinary charge). If the particle starts out with zero charge and ends up with zero charge, all is well.

There are many chargeless, or electrically "neutral," particles in nature. But the most common of all is the photon. "Consequently, it is a characteristic of Holdom's force that it would be most noticeable between photons and mirror photons," says Foot.

But how could a force between photons and mirror photons affect the lifetime of ortho-positronium? It consists, after all, of an electron and a positron. It all has to do with the "Heisenberg uncertainty principle." This allows particles to change into other particles even if the process is forbidden by the laws of physics—for instance, if the final particles have more energy than the initial ones. The catch is that the extra energy is on temporary loan.* Within a short time, it must be paid back, with the particles reverting to the way they were. It's a bit like borrowing your dad's car for the night but getting it back the next morning before he wakes up to see it gone. Particles that manage this cheeky trick are known as "virtual" particles, in recognition of their fleeting existence.†

What has this got to do with ortho-positronium? The electron and positron of ortho-positronium would desperately like to turn into a single photon. However, this is forbidden. In this case, it is not that the electron and positron do not have enough energy to make a photon; they do not have enough momentum. The total momentum of the electron and positron is zero (if the ortho-positronium is not moving),

---

*   The energy is borrowed from the "quantum vacuum."

†   All force-carrying particles are "virtual." The greater the energy borrowed to make them, the sooner it must be paid back. This leads to a connection between the range of a force and the mass of its force-carrying particles. If they take a lot of energy to make—that is, if they are very massive—they won't get far before nature notices, so the force will have a short range. An example is the weak nuclear force, which is carried by virtual particles almost a hundred times larger than a proton. The electromagnetic force, on the other hand, is carried by virtual photons, which are massless and can travel an infinite distance before nature notices them. The force therefore has an infinite range.

whereas the momentum of a photon is always nonzero. The principle, however, is the same as before. The particles can borrow the necessary momentum, as long as they pay it back within a short time.

The upshot is that ortho-positronium can briefly change into a virtual photon. The significance of this is that photons feel Holdom's force. Consequently, ortho-positronium is peculiarly sensitive to the mirror world, a fact that was pointed out by the Nobel Prize–winning physicist Sheldon Glashow of Harvard University in 1986. Specifically, before a virtual photon changes back into an electron and a positron, it has a small chance of spitting out a virtual H particle and becoming a virtual mirror photon. This can then change into a mirror electron and a mirror positron — mirror ortho-positronium.

So the ortho-positronium created in experiments is not actually pure ortho-positronium. Instead, it is a curious mixture of ordinary and mirror matter. In exactly the same way that an atom can be in two places at once, the system can be ortho-positronium and mirror ortho-positronium at the same time. It oscillates back and forth.* One moment it is 100 percent ortho-positronium, the next 50 percent ortho-positronium and 50 percent mirror ortho-positronium, the next 100 percent mirror ortho-positronium, and so on.

Now, finally, we come to the point — how the existence of a mirror universe can shorten the lifetime of ortho-positronium. According to Gninenko and Foot, if ortho-positronium is indeed flipping back and forth between the ordinary and the mirror world, then the mirror universe gives ortho-positronium another possible way of vanishing from an experiment. In addition to decaying into three photons in the standard way, it can also oscillate into mirror ortho-positronium. You might think that this would make no difference whatsoever. After all, mirror ortho-positronium promptly oscillates back into ortho-positronium. However, it turns out that there is something else significant that can happen. While in the mirror world, the system can decay into three mirror photons. If this happens, there is no longer any mirror ortho-positronium to oscillate back into the ordinary world.

---

* Recall that superpositions, in which a microscopic system is in two different states at one time — the equivalent of standing and sitting at the same time — arise because microscopic particles have a wave character and obey a "wave equation." It is a striking feature of such an equation that if it permits the existence of two individual waves, it also permits the existence of a combination, or "superposition," of the two waves. This can lead to "mixed matter," which is ortho-positronium and mirror ortho-positronium at the same time. The oscillation between the two states occurs because each wave periodically suppresses, or modulates, the other wave.

In the Michigan experiment, the physicists observed a large number of ortho-positroniums and in a given time saw more vanish than they had expected. "The reason could be that some disappeared forever into the mirror world," says Foot. "Not realizing this, the experimenters deduced a shorter lifetime."

If the oscillation into the mirror world takes a long time compared to the 142 billionths of a second it takes ortho-positronium to decay, then there will be time for only a small fraction of an oscillation before all the ortho-positronium disappears. Consequently, the effect on the lifetime of ortho-positronium will be small, as observed. According to Gninenko and Foot, ortho-positronium will have a lifetime 0.1 percent shorter than expected if it oscillates into mirror ortho-positronium and back again once every 3000 billionths of a second. It's an astonishing claim. However, the two physicists believe it might actually be testable with a new experiment.

The key is to observe the decay of ortho-positronium in an empty vacuum. This is crucial because if a single particle of matter happened to bump into ortho-positronium, it could destroy the oscillations.* According to Gninenko, this may explain why an experiment carried out by physicists at the University of Tokyo in 1995 did not observe the shortened lifetime. "There was simply too much extraneous matter around," says Foot.

Gninenko has proposed a new experiment that avoids this difficulty. The idea is to enclose ortho-positronium inside a container whose energy content can be precisely monitored. If there really is a mirror world, and ortho-positronium oscillates into mirror ortho-positronium, then stuff will disappear forever from our world. "The idea is to look for missing energy," says Gninenko. "Missing energy is the unmistakable signature of the mirror universe."

Implications of a Mirror World

According to Gninenko, his experiment could easily be done in the next few years. If it gets the go-ahead from CERN and proves that there

---

* Superpositions can exist only if the rest of the world has no knowledge of them. The instant the outside world discovers the secret—and this can happen even if a single particle of light bounces off the system and thereby takes news of it away—the system reverts to being in one state and one state only. The process that destroys quantum weirdness is called "decoherence" (see chapter 2).

is indeed an invisible mirror universe occupying the same space as the ordinary visible universe, what will it mean? Well, for one thing, it could help solve one of the outstanding puzzles of astronomy.

Over the past few decades, astronomers have come to the embarrassing realization that at least 90 percent of the matter in the universe is in some mysterious, invisible form, dubbed "dark matter." This matter reveals its presence only because it affects the motion of stars and galaxies through its gravitational pull. Nobody knows what the dark matter is. "But mirror matter is as good a candidate as any," says Foot. "As far as we know, all forms of matter, including mirror matter, exert a gravitational pull."

But if dark matter is to be explained, we need to find ten times more matter than is presently known. Surely mirror matter merely doubles the amount of matter in the universe? Not necessarily, says Foot. "Even if the fundamental laws of physics are symmetrical between ordinary and mirror matter, this does not necessarily mean that there must be equal amounts of ordinary and mirror matter," he says. "For reasons we do not yet understand, the universe may simply have started out with ten times more mirror matter than ordinary matter."

The amazing possibility of a mirror universe occupying ours raises the possibility of mirror galaxies, mirror stars, and mirror planets. You might imagine we have absolutely no chance of detecting them because they are inherently invisible. However, this may not be so. According to Foot, mirror stars might show up when they explode as mirror supernovae. A mirror supernova would produce a huge burst of mirror neutrinos. And there is reason to believe that mirror neutrinos will oscillate strongly into ordinary neutrinos. "The large flux of ordinary neutrinos could reveal a mirror supernova to underground experiments just as a burst of neutrinos revealed an ordinary supernova in 1987," says Foot.*

The mirror universe could also solve another neutrino mystery. For decades, physicists have been puzzled by curious shortfalls in the numbers of solar and atmospheric neutrinos. A strong possibility is that the shortfalls are caused by oscillations among the three known types of neutrino—the electron-, muon-, and tau-neutrinos. However, there are hints that the neutrino puzzles may require the existence of a fourth, "sterile" neutrino, so slippery that it makes the other three neutrinos

---

* This was Supernova 1987A, the first supernova to go off in our galaxy—strictly speaking, it exploded in a satellite of our galaxy—since 1604.

appear positively sociable. "One possibility is that the fourth neutrino is from the mirror universe," says Foot. "If neutrinos can oscillate into mirror neutrinos, then this can nicely explain the solar and atmospheric neutrino anomalies."

If mirror matter exists in our Milky Way galaxy, says Foot, it follows that there should exist binary star systems that consist of both ordinary and mirror matter. These would have formed from diffuse clouds of gas that shrank under their own gravity. "It is unlikely, however, that such systems would contain equal amounts of ordinary and mirror matter since the two barely interact and so are unlikely to shrink at the same rate," says Foot. "The most probable thing is that they would contain predominantly ordinary matter with a small amount of mirror matter, or vice versa."

Remarkably, claims Foot, there is tentative evidence for the existence of such systems. In the past few years, more than fifty "extrasolar" planets have been discovered orbiting nearby suns. These cannot be seen directly. However, they reveal their presence because their gravity tugs periodically on their parent stars, and this effect is just discernible from Earth. One of the most surprising discoveries is that of a class of giant planets like our own Jupiter, which orbit extremely close to their stars. One is twenty times closer than Earth is to the Sun; it is eight times closer even than Mercury, the innermost planet in our solar system.

At this tiny distance from the inferno of a star it is far too hot for ordinary planets to form. Theorists speculate that "close-in Jupiters" instead form far from the star, where the temperature is much lower, and then migrate inward. While this is possible, says Foot, there are difficulties. For example, in January 2001, a team led by Geoffrey Marcy of the University of California at Berkeley announced the discovery of a pair of planets with closely connected, or "resonant," orbits. However, migrating planets speed up as they get closer to a star, causing them to move away from one another. An intriguing alternative possibility presents itself in the mirror-world idea. "The close-in planets may be mirror worlds composed predominantly of mirror matter!" says Foot. "Such planets could actually have formed close to the star. This is not a problem for mirror worlds since they stop hardly any of the light and heat of the star and so are not significantly heated."

If the close-in planets really are mirror worlds orbiting ordinary stars, then it is natural to expect the existence of ordinary worlds orbiting mirror stars. Such planets would appear to be unaccompanied by any star. "Isolated" planets were in fact identified in the Sigma Orionis star cluster by astronomers in 2000. Such planets pose a problem for theorists because, according to the conventional picture, planets form only in dense disks of gas and dust swirling around newborn stars. "However, apparently isolated planets are quite natural from the mirror-world perspective," says Foot. "They may not be isolated at all. They could be orbiting invisible stars!"

If Foot is right, careful observations of the isolated planets might reveal periodic changes, or "Doppler shifts," in their light, the telltale sign that they are orbiting a hidden object. The astronomers who discovered the isolated planets say they have not yet been observed for long enough or accurately enough to reveal any orbital motion.

Another mind-blowing possibility is that there might be mirror planets orbiting our Sun. Foot thinks it unlikely but not impossible. The only way we could have missed such planets is if they are so small or so far from the Sun that their gravitational pull on the visible planets is too weak for us to have noticed. "Another possibility is that mirror planets could orbit in a different 'plane' to the ordinary planets," says Foot. "Such planets would be very hard to detect and we could have missed them."

Mirror planets, however, are not the only mirror possibility for our solar system. In the 1980s, it was seriously suggested that our Sun is not alone but has an ultrafaint companion star. The reason for this outlandish proposal was a hotly disputed claim that mass extinctions of life on Earth occur roughly every twenty-six million years. If the Sun's hidden companion, dubbed "Nemesis," was in a highly elongated orbit that brought it close-in every twenty-six million years, then every twenty-six million years it would stir up the Oort Cloud of comets, which surrounds the solar system. Icy bodies would be sent on a collision course with Earth and the resulting devastation could cause a massive extinction event.

Systematic searches of the sky have revealed no faint companion of the Sun. However, Zurab Silagadze of the Budker Institute in Novosibirsk, Russia, has proposed that Nemesis may exist but that it is a mirror star. "It would be very hard to prove but it is certainly fun to speculate," says Foot.

Moving down the astronomical mass scale, what about the possibility of mirror asteroids and mirror comets? Foot thinks these are certainly not out of the question. He even speculates that the body that devastated the Tunguska region of Siberia in 1908, flattening two thousand square kilometers of forest but leaving no crater, could have been a mirror asteroid or mirror comet. Because of the small interaction between ordinary matter and mirror matter, there would be only a small frictional force slowing the body and dumping energy in the atmosphere. Consequently, this event would have released relatively little energy above ground.

Ray Volkas, a colleague of Foot's at the University of Melbourne, says the idea could actually be tested by taking samples under the Tunguska site to look for evidence of a large release of energy underground. "Of course, Tunguska may have nothing to do with mirror matter," says Foot. "But I have a vivid imagination."

John Cramer is a physicist at the University of Washington in Seattle. In his 1986 novel *Twistor*, he speculated on the existence of a mirror Earth occupying the same volume of space as the ordinary Earth. The spin of the two Earths was locked so that they rotated together, and their central densities were low enough that they added up to Earth's observed central density. "It's an amazing possibility," says Foot. "But not very likely."

So, what about the *X-Files*-like possibility of mirror people walking among us on Earth? Certainly, the puzzle that we have seen no sign of extraterrestrials could be explained if they are out there but in the mirror universe. Foot points out, however, that mirror extraterrestrials coming to Earth would face a problem. We are prevented from sinking into the ground because the electrons in atoms in our feet fiercely repel the electrons in the atoms of Earth. In other words, we owe our ability to walk on Earth's surface to the electromagnetic force. What then of mirror people, who would not experience the standard electromagnetic force? "If they ever came here, they would have quickly fallen to the center of the Earth," says Foot.

The idea that our universe could, without our knowledge, be superimposed on another, "mirror" universe with its own light and matter and even stars and planets and animal life is an amazing concept. It seems astonishing that we could have overlooked such a major component of our universe. This prompts the obvious question: is there anything else we could have missed? The answer, according to a young Swedish physicist, is an emphatic yes. If Max Tegmark is right, all we see and all we surmise out to the farthest limits probed by telescopes is but a vanishingly small portion of all there is. Our universe is just one among an infinity of others, each ruled by different laws of physics.

8

The Universe Next Door

*Brace yourself. The universe is about to get bigger
than you ever imagined*

As we look out into the Universe and identify the many accidents
of physics and astronomy that have worked together to our benefit,
it almost seems as if the Universe must have known that we were
coming.

—Freeman Dyson

And I say to any man or woman, Let your soul stand cool and
composed before a million universes.

—Walt Whitman, "Song of Myself"

For much of history, people thought that the Sun and planets were
all there was to the universe—apart, that is, from a few pointlike
stars nailed to a crystal sphere. Much later, at the dawn of the twentieth
century, astronomers, aided by a new generation of giant telescopes,
discovered that the Sun was merely one among a hundred billion or so
other suns swirling in a great whirlpool of stars called the Milky Way.
Later still, and here we are talking about a mere seventy-odd years ago,
it was discovered that the Milky Way was but one "galaxy" among ten
billion others fleeing from one another like so many pieces of cosmic
shrapnel in the aftermath of a titanic explosion: the Big Bang.

"The history of astronomy," as the novelist Martin Amis has aptly remarked, "is the history of increasing humiliation." For a century now,
the human race has been shrinking, its importance in the cosmic
scheme of things dwindling as the universe grows, not in small, man-
ageable steps but in sickening lurches of the human mind. And if you
think that today, finally, we have reached the end of the road, prepare

yourself for another body blow. More and more physicists are convinced that our universe is not alone but merely one among an infinity
of others, drifting like bubbles on the river of time.

Among those who think this is physicist Max Tegmark of the University of Pennsylvania.* Tegmark imagines a "multiverse" in which the
individual universes dance to the tune of different laws of physics.

It's a remarkable idea, and Tegmark has arrived at it by a remarkable
route: by pondering an abstract and esoteric question. Why is mathematics so damned good at describing our universe?

Three and a half centuries ago, Isaac Newton discovered that the
manner in which massive bodies move when they are subjected to
forces can be perfectly described by simple mathematical equations.
Following in his footsteps, generations of physicists have appealed to
mathematics for a compact description of the world. Their spectacular
success in penetrating nature's inner workings leaves little doubt that, as
Galileo observed, "Nature's great book is written in mathematical sym-
bols." The strong implication is that God is a mathematician (!) and if
physicists ever succeed in finding a complete description of the universe—one that neatly summarizes all fundamental features of reality—
such a "theory of everything" will be a mathematical theory.

We can phrase this a little more precisely. The edifice of mathematics is built from what mathematicians call "formal systems." Familiar
examples of formal systems are arithmetic and flat-paper geometry,
sometimes known as Euclidean geometry. But mathematicians know
of many others, such as Boolean algebra and group theory. A formal
system consists of a set of givens, or axioms, and the consequences, or
theorems, that can be deduced from them by applying the rules of
logic. For instance, the axioms of Euclidean geometry include the
statement that "parallel lines never meet," while the theorems that can
be deduced from the axioms include such statements as "the internal
angles of a triangle always add up to 180 degrees."

Tegmark, who is a Swede and therefore happiest when surrounded
by pine trees, likes to think of mathematics as a towering Christmas
tree with shiny boxes hanging from all of its branches. In each box is a
different formal system. Some boxes have already been opened and
their contents examined, but many have not, providing continuing em-
ployment for the world's mathematicians. Nobody knows what won-

---

*   We met Max Tegmark in chapter 2. (Though he is young, he gets about.)

ders remain to be unboxed. "However, one thing can be said for sure," says Tegmark. "One of the unopened boxes must contain the theory of everything."

This prompts a serious question. Why should this box correspond to our universe? After all, every box on the tree of mathematics contains a formal system and, apart from obvious differences in their complexity, there is nothing to distinguish one box from another. "Why should one box be privileged above all others?" says Tegmark. "Why should one mathematical structure, and only one, out of all the countless mathematical structures be endowed with physical existence?"

Try as he might, Tegmark has been unable to answer this question. Rather than seeing this as a failure, however, his stroke of genius is to see it as a virtue. Perhaps there is a good reason why there appears to be nothing special about the box containing the theory of everything. Perhaps that reason is that there is nothing special about it.

The Dull Unexceptional Nature of the Human Race

That there is nothing special about a circumstance, most notably the position of Earth in the universe, has proved to be a powerful guiding principle in science. In the sixteenth century, the Polish astronomer Nicholas Copernicus maintained that movement of the Sun and planets across the sky could best be explained if Earth were not, as most people supposed, the central pivot about which the cosmos turned, but if instead it were just another planet circling the Sun.

In the twentieth century, astronomers extended the "Copernican principle" to the wider cosmos revealed by their giant telescopes. Just as Copernicus had maintained that there was nothing special about the place of Earth in what we now know as the solar system, so modern astronomers maintained that there was nothing special about the place in the universe of the Milky Way, the galaxy that contains Earth and the Sun. On this apparently flimsy bedrock—the dull, unexceptional nature of the Milky Way—is founded the entire edifice of "cosmology." Cosmology is the science that has taught us that we live in a universe that is expanding and cooling in the aftermath of a violent explosion that took place twelve to fourteen billion years ago.

Tegmark's brainstorm is to take the tried-and-tested Copernican principle and extend it to the tree of mathematics. It is his contention that there is absolutely nothing special about the box that contains the theory of everything—nor, for that matter, about any of the other boxes on the tree of mathematics. Every single one has equivalent status. All, in other words, correspond to universes.

Pause for a moment to take in what this means. Tegmark is saying that, in addition to a universe that dances to the tune of the theory of everything—the one we find ourselves living in—there is a universe governed by the laws of arithmetic, a universe ruled by the laws of two-dimensional geometry, and so on, ad infinitum. Out there in the big wide multiverse, perhaps beyond the farthest limits yet probed by our telescopes, there are universes that dance to the tune of all conceivable mathematical equations.

Consider the implications of this audacious idea. In the 1930s, the Austrian physicist Eugene Wigner famously remarked on "the unreasonable effectiveness of mathematics in the physical sciences." Nobody has come up with a satisfactory explanation of why it is so effective at encapsulating the essence of the universe. No doubt even Newton in his time puzzled over the matter. But, if Tegmark is right, suddenly it is abundantly clear why mathematics is so good at describing physics. The answer is almost laughably trivial. Mathematics is so good at describing physics because mathematics is physics.

According to Tegmark, every last formal system on the tree of mathematics is endowed with physical existence. Every last formal system corresponds to an actual universe.

Tegmark has extended the Copernican principle beyond our universe to an infinite set of universes. Not only is there nothing special about our status within the universe, he claims; there is nothing special about the status of our universe within the infinity of universes that constitutes the multiverse.

The Evidence for Other Universes

The idea that there exists a multiplicity of universes is not new. Tegmark has merely picked it up and taken it to its dizzy extreme. The most striking evidence that there is a multiverse comes from the fundamental laws that control the universe. A peculiar thing about these laws is that they appear to be fine-tuned so that human beings, or at least living things, can exist.

This remarkable fact was first noticed by the British astronomer Sir Fred Hoyle. In the 1950s, Hoyle discovered that the step-by-step process by which heavy atoms are built up from light ones deep in the central furnaces of stars depends on a series of odd nuclear coincidences. Only if the cores, or nuclei, of three atoms—beryllium-8, carbon-12, and oxygen-16—have just the right energy can hydrogen, the lightest atom, be assembled into heavy atoms such as calcium, magnesium, and iron, which are the essential ingredients of life.*

If Hoyle's example of nature's fine-tuning was the only one, it might be possible to sweep it under the carpet. However, it isn't. "Many instances have been found in which if a certain fundamental force of nature were slightly weaker or stronger, or if a certain fundamental particle were slightly lighter or heavier, there would be no galaxies or stars or planets, and hence no human beings," says Tegmark.

Take gravity, for instance. If the force of gravity were just a few percent weaker than it is, it would be quite incapable of squeezing and heating the matter in the hearts of stars to the many millions of degrees necessary to trigger the nuclear reactions that generate sunlight. Stars such as the Sun would consequently not exist. If, on the other hand, gravity were only a few percent stronger than it is, it would boost the temperature in the cores of stars, causing them to consume their fuel faster and burn out more quickly. Though stars would still exist, they would not exist for the billions of years needed for biological evolution to produce intelligent life.

In addition to gravity, there is the strong nuclear force, which is responsible for gluing together atomic nuclei. If the strong force were just a few percent stronger, the Sun would burn its entire supply of hydrogen fuel in less than one second and consequently explode. Instead, the Sun uses up its fuel at a leisurely rate over about ten billion years, a period of time far more suited to the evolution of intelligent life.[†]

---

* The atomic nucleus is the dense clump of matter at the center of an atom. It is made of two main constituent parts—protons and neutrons (the exception being the nucleus of hydrogen, the simplest atom, which consists of a lone proton). The "8" in beryllium-8 refers to the total number of building blocks in its nucleus.

† Isaac Asimov explored what atoms, stars, and life would be like in a universe where the strong nuclear force was stronger in his award-winning novel *The Gods Themselves*. The central character discovers the existence of such a universe when he stumbles over a sample of plutonium-186. In our universe, plutonium needs the "glue" from at least 240 protons and neutrons to hold itself together, and would explode if it had as few as 186 particles.

If the strong force were a few percent weaker than it is, on the other hand, it would be too feeble to hold together deuterium, an essential step in the generation of starlight and the first stepping-stone in the building of atoms heavier than hydrogen inside stars.* The universe would therefore lack the heavy atoms that are essential for life.

In addition to the strong force, there is a second force that operates in the domain of the atomic nucleus. And, surprise, surprise, it turns out that it too is finely balanced so that we can exist.

The weak force plays a crucial role in the explosion of massive stars as supernovae. Specifically, it is the means by which subatomic particles known as neutrinos interact with matter. Neutrinos are created in vast numbers deep in the core of a dying star and, as they surge outward into space, they are instrumental in blowing away the outer layers of the dying star.

If the weak nuclear force were only a smidgen stronger than it is, the neutrinos would interact so much with matter that they would be stopped dead in the star's core. With no neutrinos left to drive away the outer layers, the explosion would stall well before it could rip apart the star. If the weak nuclear force were slightly weaker than it is, on the other hand, the neutrinos would interact so little with matter that they would escape into space without interacting much at all with the material of the star. Once again, there would be nothing to blow away the star's outer layers and create a supernova.

What has this got to do with us? Well, the heavy atoms that are essential for life are forged in the furnaces of massive stars. And those heavy atoms remain locked away forever unless such stars go supernova and catapult them into space. Consequently, if the weak force were just a fraction weaker or a fraction stronger than it is, no iron or calcium or iodine or many other atoms essential for life could have been blown into space to be incorporated into new stars and planets and, ultimately, you and me. "Wherever physicists look they see examples of fine-tuning," says Astronomer Royal Sir Martin Rees of the University of Cambridge. "Most of the physical constants and the initial conditions of the universe examined so far appear to be fine-tuned to some extent."

By "physical constants" or "fundamental constants," physicists mean the quantities that ultimately govern the universe. These include such

---

* Deuterium is a heavy form of hydrogen, the lightest atom. Instead of containing a single proton in its nucleus, it contains a proton and a neutron. Combined with oxygen, deuterium makes "heavy water."

things as the numbers that characterize the strength of the four fundamental forces, or the masses of the fundamental particles.

What are we to make of the fine-tuning of the physical constants? "It could be a coincidence," says Rees. "I once thought so. But that view now seems too narrow."*

Tegmark agrees that nature's fine-tuning cannot be passed off as a mere coincidence. "There are only two possible explanations," he says. "Either the universe was designed specifically for us by a creator, or there exists a large number of universes, each with different values of the fundamental constants, and not surprisingly we find ourselves in one in which the constants have just the right values to permit galaxies, stars, and life."

"What's conventionally called 'the universe' could be just one of an ensemble," says Rees. "The universe in which we've emerged belongs to the unusual subset that permit complexity and consciousness to develop."

Among the universes of the multiverse will be universes in which the strength of gravity is weaker or stronger than ours, universes in which the strong force is weaker or stronger, and so on. There will be a universe corresponding to each and every permutation it is possible to imagine. "The concept of the 'multiverse' is potentially as drastic an enlargement of our cosmic perspective as the shift from pre-Copernican ideas to the realization that the Earth is orbiting a typical star on the edge of the Milky Way, itself just one galaxy among countless others," says Rees.[†]

But the evidence for a multiverse does not come only from the fine-tuning of the fundamental constants of physics. "It is actually very difficult to construct a theory where everything we see is all there is," says the physicist Andreas Albrecht of the University of California at Davis. "Nature is telling us from a host of different directions that our universe is only one among a very large number of other universes," claims Tegmark.

One of those other directions is string theory. This holds out the tantalizing hope of merging, or unifying, quantum theory with Einstein's theory of gravity. However, it exacts a cost. The theory views quarks and leptons not as pointlike particles but as ultratiny pieces of "string" vibrating in a space-time of ten dimensions.

The outstanding problem faced by string theorists is not too difficult to guess. Why, if ultimate reality is ten-dimensional, do we human

---

* Martin Rees, *Before the Beginning* (London: Simon & Schuster, 1997).
† Ibid.

beings experience a mere four dimensions—three of space and one of time?

The standard answer from the string theorists is that, for some reason, six of the ten dimensions must be rolled up into small loops. The loops must be extremely tiny—far smaller even than an atom—or else we would have already noticed them. However, if the extra dimensions are indeed rolled up then this poses another problem. Why should precisely six dimensions be rolled up? Why not one or five or nine? String theory, unfortunately, is silent on this question.

But Tegmark takes a leaf out of the book of the Nobel Prize–winning physicist Steven Weinberg. "Our mistake," Weinberg has written, "is not that we take our theories too seriously but that we do not take them seriously enough."

According to Tegmark, the reason string theory does not pick out a universe with six rolled-up dimensions as in any way special is simply that there is nothing special about such a universe. It's the Copernican principle again. String theory tells us that there is an ensemble of universes with all possible permutations of rolled-up dimensions. Say the dimensions are labeled from one to ten. Then there is a universe with dimensions five to ten rolled up small; another universe with dimension one rolled up; another with dimensions two to ten rolled up; and so on. "There are literally thousands upon thousands of possible universes corresponding to all the possible ways of rolling up the dimensions," claims Tegmark.

So why then are we in the universe with four big dimensions and six rolled-up ones? To find out, Tegmark has systematically examined what physics would be like in universes that do not have three spatial dimensions and one temporal dimension. He finds that in universes with fewer than three spatial dimensions there is no gravity and there are serious "topological problems." "For instance, in two-dimensional space, one problem is that nerve fibers cannot cross," he says. "Another is that a cat that eats something, then excretes it, is cut in half. It's the kind of thing that was explored more than a century ago by Edwin Abbott in his classic mathematical fable *Flatland*."

Tegmark has also investigated universes with more than one temporal dimension. Contrary to expectations, he says, they are not too weird to think about, because any observer's perception of time could remain one-dimensional. "The problem is that physics is infinitely sensitive to initial conditions so it's impossible to predict the future," says Tegmark.

In universes with more than three spatial dimensions, a problem rears its head; it was first pointed out by the physicist Paul Ehrenfest in 1917. There are no stable orbits either in classical or in quantum physics. Particles, be they electrons or planets, either spiral together or fly off to infinity. "Not only does this rule out the existence of planetary systems like our own but it also rules out the existence of atoms," says Tegmark.

Tegmark concludes that only in a universe of three spatial dimensions and one temporal dimension is physics likely to provide three things—the richness, predictability, and stability necessary to generate interesting phenomena such as life. "I can't say categorically that other space-times cannot contain observers like us," he says. "But I'd say the prospects are pretty bleak."

Tegmark's key discovery is that, if you assume there are other universes with different numbers of dimensions, for the first time it is possible to understand why we live in a universe with three dimensions of space and one of time. He sees this as evidence that the multiverse contains not only universes with all possible permutations of the fundamental constants but also universes with all possible permutations of space-time dimensions.

But why stop here? Why indeed? And this is where Tegmark goes way, way beyond everyone else who is willing to countenance the idea of a multiverse.

The Ultimate Multiverse

According to Tegmark, universes governed by different fundamental constants and different numbers of large dimensions make up only a tiny fraction of the universes that exist in the extraordinary multiplicity that is the multiverse. Think of vehicles. Cars with different designs or colors are only a tiny subset of all possible vehicles. In addition, for instance, there are bicycles and bullet trains, nuclear submarines and moon rockets. Similarly, Tegmark maintains that universes with different fundamental constants or numbers of large dimensions make up only a small subset of universes governed by different equations of physics.

The set of universes corresponding to all possible different mathematical equations includes all other conceivable ensembles of universes. Tegmark therefore calls it the "ultimate ensemble." He got the idea of the ultimate ensemble while a graduate student at the University of California at Berkeley in 1992. "At first, I suggested it to a friend as a joke," he says. "But the idea kept coming back—I just couldn't stop thinking about it."

In the summer of 1996, Tegmark wrote up the idea in a scientific paper. For a while it remained unpublished because Tegmark could not think of a physics journal where it might fit. Instead, he posted it on the Internet, where it generated much interest. There is even an Internet newsgroup set up by physicists and computer scientists to discuss the paper and its implications.*

So, where are all these other universes? After all, when astronomers look outward from Earth with their telescopes, they see only one universe. Where is the evidence of a multiplicity of other universes? The short answer is that neither Tegmark nor anybody else knows. However, one possibility is provided by the popular theory of "inflation," which purports to describe the first split second of the universe's existence. If inflation is correct, the observable universe is merely a tiny, tiny bubble in a tremendously bigger universe. Some theorists have even suggested that the laws of physics "freeze out" differently in different bubbles, which might conceivably explain the location of some of Tegmark's extra universes.†

One question that springs to mind is: how does all this connect with the concept of the Many Worlds? Chapter 2 presented the idea that there may exist multiple realities, or universes, in which all possible histories are played out. Here the idea is that there are multiple universes constructed according to all possible laws of physics.

Tegmark is well aware of the similarity. "Nature is telling us from many different directions that there is more than one universe/reality," says Tegmark. "However, at the moment, we are simply not at the stage where we can see how the pieces fit together to create the big picture."

So, is there a payoff for believing in a large ensemble of universes? The answer, according to Tegmark, is an emphatic yes. Admitting that

---

* See the "everything mailing list" create by Wei Dai (e-mail address: everything-list@ eskimo.com; website: http://escribe.com/science/theory).

† As yet, there is no fundamental theory that explicitly predicts that the laws of physics, including the fundamental physical constants, could vary in the way required. This is the Achilles' heel of the multiverse idea. On the other hand, however, there is as yet no fundamental theory that pins down the laws of physics to the ones we observe. A betting man would not rule out either possibility at this stage.

there is a multiplicity of universes in which anything goes may actually have predictive power. Take the hypothetical theory of everything, which completely describes our universe. According to the American physicist John Wheeler, famous for coining the term "black hole," even if physicists one day get their hands on a theory of everything, they will be disappointed. Why? Because they will still face an unanswerable question: why does nature obey this set of equations and not another?

According to Tegmark, having an infinity of universes offers a way out of this dilemma. How? Well, the universe we are in—and hence the theory of everything—is determined by our presence. The conditions must be right for human beings to evolve. This should select our universe from the infinity of possible universes and tell us why the theory of everything has the precise form it does. Biology, it seems, ultimately determines physics.

We're Here Because We're Here Because . . .

The idea that the universe is the way it is because we are here is known as the "anthropic principle." Its curiously topsy-turvy logic was used at the outset of this chapter to deduce from the fine-tuning of physics the existence of a multiverse. Tegmark also used it to come up with a possible explanation for why our universe has four big space-time dimensions. And it can also be used in the case of the ultimate ensemble—an infinity of universes, each ruled by different laws of physics. "The key thing is that although every mathematical structure exists and has physical existence, only some are perceived to have physical existence," says Tegmark. "For instance, a universe consisting of Euclidean geometry exists but its axioms are nowhere near rich enough in possibilities to evolve intelligent life."

Intelligent life will evolve only in universes where the conditions are right for intelligent life to evolve. The reason the universe is the way it is is that, if it weren't, we would not be here to notice.

The key thing, says Tegmark, is to define precisely the conditions necessary for life. Armed with this definition, he maintains, it will be possible to sift through the multiverse and find those universes that fit the bill. This is likely to be a tough task. "You're trying to get precision in physical science by invoking just about the most imprecisely defined thing—life," says Albrecht.

When Tegmark talks about the "conditions necessary for life," he is careful not to exclude the possibility of nonbiological life. After all, for all we know, it may be possible to have life based on the element silicon, rather than carbon, or life based on computer software, or on some foundation as yet undreamed of. Tegmark has coined the slightly cumbersome term "self-aware substructures," or SASs, to represent all possible forms of life. The conditions necessary for the evolution of SASs constrain the theory of everything. "Biology in a wider sense than we normally use it determines the laws of physics," says Tegmark.

The task of finding the theory of everything therefore boils down to defining the conditions necessary for SASs so that those conditions can be used to pick out universes from the multiverse. Tegmark thinks it unlikely that there will turn out to be just one universe capable of supporting SASs. "Rather than an island in parameter space, I think there is an archipelago," he says.

Some of the universes will have equations of physics close to ours, others more variant, still others even more variant, and so on. But Tegmark makes a key prediction on which the whole ultimate ensemble idea stands or falls: the most common universes will be the ones with equations of physics very close to ours.

This is nothing but the Copernican principle once again. Our universe must turn out to be a typical, run-of-the mill universe among the universes capable of supporting SASs. There must be nothing whatsoever special about it. "If there is, then I am wrong," says Tegmark. "It's the definitive test of my ultimate ensemble idea."

The Wasteful Multiverse

The stark reality of the ultimate ensemble—indeed, of any version of the multiverse—is that most universes have conditions that leave them uninteresting and devoid of life. Many people find this abhorrent, and many scientists oppose the idea. "It's an emotional reaction—a gut feeling," says Tegmark. "They just don't like the waste of all those dead universes."

But Tegmark thinks this is not a scientifically valid objection. "Reality is under no obligation to be palatable to human beings," he says. "We have to put up with whatever it throws at us."

Tegmark has an extraordinary argument with which to counter his critics. He claims that there may actually be less information in the multiverse than in an individual universe. The whole, in other words, may be simpler and less wasteful than any of its parts.

To illustrate his argument, Tegmark considers the whole numbers, or integers. A useful measure of the complexity, or information content, of a number is the length (in binary digits—"0"s and "1"s) of the shortest computer program that will produce that number as output. By this measure, the information content of a randomly chosen, or generic, integer n is of order $\log_2 n$. (If this is too technical, don't worry. The point is that the information content of a generic integer may be quite high.) By contrast, however, it is possible to generate the set of all integers 1, 2, 3, . . . by quite a trivial computer program which starts with 1, adds 1, and loops back to do the same thing. "Here, the complexity of the whole set is smaller than that of one of its parts," says Tegmark. "And my guess is that, paradoxically, the multiverse contains less information than any single universe."

But what if you don't buy Tegmark's argument that there isn't any waste because the multiverse contains less information than individual universes? What if you think it's all fancy footwork, sleight of hand? Then, surely, you have no choice but to accept the idea of a multiverse with endless dead universes. It would certainly seem so. However, one physicist believes there is a way to have a multiverse but avoid the terrible waste. His name is Ed Harrison and, if he is right, we may at last find out where the universe came from. Simple, says Harrison. It was created by superintelligent beings living in another universe.

9

Was the Universe
Created by Angels?

*The discovery that it might be possible to make a universe
in the laboratory could have profound implications
for the origin of our own universe*

Since the inflationary theory implies that the entire observed universe can evolve from a tiny speck, it is hard to stop oneself from asking whether a universe can in principle be created in the laboratory.
—Alan Guth, *The Inflationary Universe*

Teacher told my parents that I am the slowest youngster in my class, but today I made a star in the third quadrant of kindergarten.
—James E. Gunn, "Kindergarten"

The ultimate experiment is about to begin. On a cold, lonely moon, shrouded in purple-pink fog, a sentient ocean marshals the energy resources of an entire galaxy and focuses them down onto a tiny, unsuspecting mote of matter. A hundred billion stars flicker and dim. The air above the ocean sizzles and catches fire. Crushed by stupendous energies, the tiny mote twists and bucks and, with a violent shudder, implodes like a nuclear explosion in reverse. Smaller and smaller it shrinks. Smaller than an atom. Smaller than the smallest subcomponent of an atom. On and on into submicroscopic realms beyond human imagination. Until, suddenly, without warning—puff!—it is gone.

Somewhere else—in another space, another time—a searing-hot fireball explodes out of nothingness and immediately begins to expand and cool. The ultimate scientific experiment has produced the ultimate experimental result: the birth of an entirely new universe.

Could our own universe have been born in such an experiment? One man thinks it is a real possibility. According to Edward Harrison, formerly of the University of Massachusetts at Amherst, our universe could easily be the outcome of an experiment carried out by a superior intelligence in another universe.

Why would anyone suggest such an outlandish idea? Because it can potentially shed light on a deeply puzzling feature of our universe. The puzzle, highlighted in the previous chapter, concerns the laws of physics that orchestrate the life of the cosmos. Physicists have discovered that even slight deviations in the laws that we observe would result in a universe completely devoid of stars and life.

What are we to make of this "fine-tuning" of the laws of physics?* There would appear to be only two possible explanations. One is that the universe was designed specifically for us by God, a Supreme Being. The other is that the universe is the way it is because if it weren't we would not be here to note that. According to this curiously topsy-turvy reasoning, known as the "anthropic principle," it is hardly surprising that we find ourselves in a universe that is fine-tuned for the existence of galaxies, stars, and life. We could hardly have evolved in a universe that was not.

The anthropic principle leads naturally to the idea that our universe is not alone but instead part of a large ensemble of universes. In each individual universe of this multiverse, the strengths of the fundamental forces take on different values, the fundamental particles have different masses, and so on. Or, to take the extreme point of view of Max Tegmark, the laws of physics are entirely different.

The possibility that our universe was designed specifically for life by a creator is something accepted by many people, including some scientists. "The drawback of this explanation, unfortunately, is that it terminates all further scientific enquiry," says Harrison.

The other logical possibility—that there exist countless other universes besides our own—is also perfectly plausible, according to Harrison. However, an unavoidable consequence of the idea is that the overwhelming majority of universes will not have the special conditions needed for the birth of galaxies, stars, planets, and so on. Harrison finds this unappealing. "The multiverse idea requires the existence of countless uninteresting and lifeless universes," he says. "To me, this is waste on a truly cosmic scale."

---

* Strictly speaking, physicists talk about the fine-tuning of the "fundamental constants" of physics. By this, they mean the strengths of nature's fundamental forces, the masses of its fundamental particles, and so on.

But, surely, if Harrison does not accept the idea of a multiworld waste-land of mostly dark and barren universes, then mustn't he accept that physics was fine-tuned by a Supreme Being beyond all rational enquiry? Not necessarily. In cosmology, as in politics, there may be a third way. According to Harrison, the multiverse could be as far from a wasteland as it is possible to imagine. It could be totally dominated by universes with galaxies and stars and life. There is only one prerequisite. Life-bearing universes must have a special ability: the ability to reproduce.

Self-Reproducing Universes

Harrison is not the first to propose such an idea. A few years ago, the physicist Lee Smolin, then at Syracuse University in New York, latched onto a speculation about what happens deep inside black holes, formed from the catastrophic shrinkage of stars.* The interiors of black holes are deeply mysterious places, forever beyond our view, where the accepted laws of physics provide no guide. This has not, however, dis-couraged physicists from speculating about what goes on there. One idea is that the shrinking interior of a black hole shrinks only so far be-fore it rebounds as another universe with slightly different laws of physics. Not in our universe, mind, because it is a law of black holes that nothing that is inside can ever get out again, but somewhere else.

If, as Smolin believes, black holes give birth to baby universes, then the universes that are geared up to produce the most black holes will spawn the most offspring universes. If the offspring universes are similar to their parents, then, inevitably, universes that make lots of black holes will come to dominate the multiverse.† It follows that we ourselves must be living in a universe optimally suited for making black holes.

But there is a snag. The prerequisite that life-bearing universes should come to dominate the multiverse is not that universes with lots of black holes should make more universes with black holes but that

---

* A black hole is a region of space where gravity is so strong that nothing, not even light, can escape. It is commonly believed to form when a massive star exhausts the fuel in its core and shrinks catastrophically under its own gravity. However, much bigger black holes—as massive as millions or even billions of suns—appear to reside in the central re-gions of most galaxies, including our own Milky Way. How these monsters formed is at pre-sent a mystery.

† Lee Smolin, *The Life of the Cosmos* (New York: Oxford University Press, 1997).

life-bearing universes should spawn more life-bearing universes. Smolin is well aware of this. He therefore argues that the same laws of physics that promote the formation of black holes must also promote the emergence of biology. Harrison, however, finds this a bit hard to swallow. "I can see no compelling reason why universes which make lots of black holes should also be good for life," he says.*

Instead, Harrison proposes a novel twist on the idea of self-reproducing universes. Life-bearing universes come to dominate the multiverse, he maintains, because intelligent life actively makes new universes. Forget black holes. Life itself takes over the universe-building business. "In offspring universes which are fit for life, new life evolves to a high level of intelligence, then creates further universes," says Harrison. "On the other hand, universes which are unfit for life evolve no life and so fail to reproduce."

In Harrison's scheme, the laws of physics that are most suited for the emergence and evolution of life are naturally selected by life itself. For this reason, he calls his idea the "natural selection of universes." If Harrison is right, then the origin of our universe is no longer such a mystery. It was created by superintelligent beings living in another universe entirely.

But how does this explain the fact that the laws of physics in our universe are fine-tuned for life? According to Harrison, there are two possibilities. The first, already touched upon, is that new universes naturally inherit the characteristics of their cosmic parents much as children inherit the characteristics of their human parents. Small "genetic variations" in the laws of physics between generations would ensure that new universes were not carbon copies of their predecessors. It follows that since the parent of our universe was fine-tuned for life and similar to our own—if it wasn't, life would never have arisen in it to make our universe—our universe must also be fine-tuned for life.

If, however, the characteristics of a parent universe are not automatically inherited by their offspring, there is another possible explanation for the fine-tuning we have observed. The makers of our universe actively engineered our universe to have laws of physics that promoted

---

* The Russian physicist Andrei Linde has proposed a self-reproducing universe that is "undirected." In his theory, called eternal, or chaotic, inflation, baby universes are constantly springing up spontaneously in a timeless "metauniverse," and giving birth to babies of their own. The Princeton physicist Richard Gott III has pointed out that a baby could beget a baby that might beget a baby that might ultimately give birth to the universe that started it all. In other words, the universe could end up being its own great-grandmother.

the evolution of intelligent life. Strictly speaking, this would not be "natural selection," the hallmark of Darwinian evolution. Natural selection occurs only if the variations—in this case, variations in the laws of physics—are random. If conscious life in parent universes engineers, or "programs," the variations, then what is happening is more like genetic engineering. According to Harrison, it should more accurately be called "self-directed selection."

If Harrison is on the right track about the natural selection of universes by intelligent life—or even the self-directed selection of universes by intelligent life—then the mystery of why the universe appears designed for the benefit of life has a deceptively straightforward solution. It appears designed for life because, at a fundamental level, it was designed for life. However, and this is the novel twist supplied by Harrison, it was designed not by God—a Supreme Being—but by superior beings. Angels, if you like. "Intelligent life takes over the business of making universes," says Harrison. "Consequently, the creation of the universe drops out of the religious sphere and becomes a subject amenable to scientific investigation."

There is a crucial assumption in Harrison's reasoning that has been quietly passed over. The assumption is that it is actually possible for a sufficiently advanced civilization to engineer a new universe. Surely this is pure science fiction? Bizarre as it seems, it is not. For more than a decade now, physicists have known—in principle if not in practice— how to trigger the birth of a new universe.

How to Build a Universe

The recipe was discovered independently by the Russian physicist Alexei Starobinsky in 1979 and the American physicist Alan Guth in 1981. Starobinsky and Guth had both been thinking about the first split second of the universe's existence and, in particular, the state of the vacuum at that time. Most people think of the vacuum as empty space, but in the eyes of modern physicists the "quantum vacuum" is a quite different beast—a roiling sea of energy that is anything but empty.

What Guth and Starobinsky realized was that in the first split second of the life of the universe, when its density was a staggering 10^{94} grams per cubic centimeter (that's 1 followed by ninety-four zeroes), the vacuum existed in a peculiar state indeed. It possessed a sort of "antigravity" that drove the universe to expand, or inflate, at a phenomenally fast rate. But this, it turns out, was the least of the vacuum's peculiar properties. Most bizarre of all was its ability to conjure energy out of nothing at all.

Normally, when anything expands—for instance, the cloud of hot debris created by the explosion of a bomb—it thins out and becomes less dense. Not so the vacuum at the beginning of the universe. Unlike anything in the everyday world around us, the vacuum expanded at a constant density and never thinned out. Imagine holding a stack of bank notes between the palms of your hands, pulling them apart and discovering that more and more bank notes materialize out of thin air so that, miraculously, the space between your palms is always filled with bank notes. It's not a very likely moneymaking scheme. However, according to Starobinsky and Guth, this was exactly how the vacuum at the beginning of time behaved. As it expanded, ever more vacuum was created. Energy simply flooded out of nothing.

Eventually, and this was still within the first split second of the universe's existence, this inflation ran out of steam. Abruptly, the enormous energy stored in the vacuum was dumped into the universe's matter, heating it to around a billion billion billion degrees. This was the searingly hot fireball we have come to call the Big Bang.*

If the inflationary picture is correct, then our universe arose from a super-dense "seed" of matter that triggered a runaway inflation of the vacuum. After this phase, which lasted a mere fraction of a second, the balance of matter—the huge amount needed to make the countless stars and galaxies we see around us—was created from the prodigious

---

* "Inflation" explains several puzzling features of our universe. For instance, if you imagine running the expansion of the universe back in time like a movie in reverse, you come to an epoch shortly after the birth of the universe when all of creation was squeezed into a volume just a millimeter across. At that epoch, the distance light could have traveled since the beginning was smaller than a millimeter. In fact, it was smaller by an enormous factor—10 followed by 31 zeroes. Now, the only way that a region of space can know about the conditions in another region of space is if some influence travels between them—and the maximum speed of any influence, according to Einstein, is the speed of light. So the millimeter-sized early universe consisted of 10^{93} regions that could not have known about each other. Here is the problem. If that millimeter-sized universe expanded to become our universe, how do we explain why the number of galaxies in a given volume of space is the same everywhere? We have to explain how 10^{93} regions that could not have known about each other got to know about each other. Inflation explains the puzzle by saying that our universe did not evolve from that millimeter-sized primordial universe. Instead, it inflated from just one of the 10^{93} regions. Consequently, beyond the "horizon" of our observable universe, there are at least another 10^{93} regions like our own.

energy of the vacuum.* The universe, as proponents of inflation are fond of saying, was the "ultimate free lunch."

That the birth of our universe could have been triggered by a tiny seed of matter greatly impressed Guth. Shortly after the discovery of inflation, it prompted him to make one of the most outrageous suggestions in the history of science. Guth suggested that it might be possible to make a universe in the laboratory.

The recipe was clear. Take a seed of matter. According to the Russian cosmologist Andrei Linde, as little as a thousandth of a gram would be enough. Next, squeeze the seed to the extraordinary density that was once sufficient to trigger the inflation of our own universe. Matter crushed to such enormous densities will form a black hole. According to Guth's theory, however, the superdense interior of such a black hole will immediately inflate—not in our universe, but in a bubble-like space-time connected to our own by the "umbilical cord" of the black hole. The umbilical cord is not stable. Tiny black holes have a habit of living for only a split second before disappearing, or "evaporating," in a sleet of so-called Hawking radiation. The moment this happens, the umbilical cord will snap and, presto, a new baby universe will be born.

The devil is in the details. Harrison, however, is not too concerned. "Precisely how a universe is made in practice is not important," he says. "The important thing is that if beings of our limited intelligence can dream up wild, yet seemingly plausible, schemes for making universes, then beings of much higher intelligence might know theoretically and technically exactly how to do it."

The Universe-Building Business

Guth's suggestion that a universe could in principle be made in the lab as an experiment was little more than a bit of fun. After all, recreating the conditions that existed in the first split second of the universe involves compressing matter to 10^{94} grams per cubic centimeter. Not only is this far beyond our current technical capabilities, it is likely to

---

* Matter, according to Einstein, is merely a compact form of energy. It can be converted into other forms of energy, such as light or heat, and other forms of energy can be converted into matter. At the end of inflation, the vacuum energy could therefore have been converted into large quantities of matter which, when it cooled, formed stars and galaxies, including our own Milky Way.

remain far beyond our capabilities for a very long time. But—and this is Harrison's point—the feat may not be entirely impossible. "It's perfectly conceivable that more-intelligent beings—perhaps even our own descendants in the far future—might possess not only the knowledge but also the technology to actually build universes," he says.

According to current estimates, our universe has existed for between about twelve and fourteen billion years. The implication is that elsewhere in the cosmos there could be technological civilizations that are millions, or even billions, of years ahead of us.* Think how far we have come in only the past century. To an inhabitant of 1900, most of present-day technology—from televisions to mobile phones to computers—would be indistinguishable from magic.† What more might we achieve if we manage to survive for another century? Or another thousand years? Perhaps it is not inconceivable that a civilization millions of years more advanced than we are might actually be able to make universes.

But why would it want to? One possibility, Harrison points out, is simply to prove that something can be done and to see what happens. Human beings often do things for no better reason. Perhaps there are some beings so advanced that their children make universes in the same way human children make figures out of plasticine. Such an idea was explored by the science fiction writer James Gunn in his story "Kindergarten."

Another possibility, says Harrison, is that an advanced civilization might make new universes out of a spirit of altruism. Our universe is clearly hospitable to intelligent life—we are here, after all. However, it may not be the most hospitable universe. Like a benevolent creator, altruistic beings might want to make universes that are even more hospitable

---

* Some scientists dispute the figure of billions of years. They point out that life like ours requires rocky planets and rocky planets are made of atoms heavier than hydrogen such as silicon and iron. Such atoms did not exist when the universe was born but have been forged since in the furnaces of stars and blown into space, where they have been incorporated in new generations of stars. The point is that it takes a long time to build up the sort of abundance of heavy elements needed to make an Earth-like planet. Life on Earth may therefore have got started at the earliest possible moment in cosmic history. Add to this that it has taken evolution more than four billion years to produce us and it could be argued that we may be among the first intelligent races to have arisen. Certainly, this is often cited as the reason our searches for radio signals from extraterrestrial civilizations have so far failed. Another view is that we simply have not been looking long enough.

† The science fiction writer Arthur C. Clarke has even stated this as one of his "three laws." "Any technology that is sufficiently advanced is indistinguishable from magic."

to intelligent life. Such a motivation was anticipated by the medieval philosopher Alphonso the Wise. "Had I been present at the Creation," he wrote around 1270, "I would have given him some useful hints for the better ordering of the universe."

It may be that advanced beings make new universes for reasons totally beyond our comprehension. Whatever their motivation, however, it is possible to speculate on the number of new universes that might be spawned. There are about ten billion galaxies like our own Milky Way in the observable universe. If, during the lifetime of each galaxy, a single civilization emerges that makes a new universe—a modest figure when you consider that our galaxy alone has two hundred billion suns—then our universe will manage to reproduce ten billion times. Furthermore, if intelligent life in each galaxy of each daughter universe repeats the ultimate experiment just once, the result is ten billion times ten billion granddaughter universes. And so on, ad infinitum. This kind of cosmic birth rate puts a flu virus to shame. It is not difficult to see how life-bearing universes could quickly come to dominate the multiverse.

Why Is the Universe Comprehensible?

Harrison's idea has real explanatory power. Not only does it explain why the laws of physics in our universe are fine-tuned for life; it also sheds light on one of the deepest puzzles in science. The puzzle was pointed out by Einstein. "The most incomprehensible thing about the universe," he said, "is that it is comprehensible."

What Einstein meant was that it is easy to imagine a universe with laws that are so complex and opaque that they are completely unfathomable by human minds. Instead, the universe appears to be governed by rather simple laws. So simple that, more than three centuries ago, Newton was able to deduce a universal law—the law of gravity. And, since Newton's time, our amazing success in penetrating nature's inner workings has given us unprecedented control over the material world. Why has it been so easy?

If Harrison is correct, the answer is simple. The reason our universe is comprehensible is that it was created by comprehensible beings. Beings far in advance of us but basically like ourselves. Intelligent but also intelligible. They made our universe to be like theirs, and their universe was in turn understandable. How could it not be? They had to have enough understanding of it to manipulate it and make a new universe.

"We have found a strange footprint on the shores of the unknown," wrote the English astronomer Arthur Eddington. "We have devised profound theories, one after another, to account for its origin. And at last we have succeeded in reconstructing the creature that made the footprint. And lo! It is our own."* Not quite. According to Harrison, the footprint was made not by us, but by beings similar but superior to us. By angels.

In Harrison's picture, life begets life, intelligence begets intelligence. "It is not inconceivable," he says, "that the goal of the evolution of intelligence is the creation of universes to foster intelligence."

But maybe the goal is more prosaic than the spreading of the miracle of life. Piling speculation upon speculation, maybe it is possible to travel between universes. "Offspring universes may not be totally inaccessible to their creators," says Harrison. "If intelligent beings know how to create universes, they might also know how to explore and occupy them."

How Did It All Begin?

Harrison's is a mind-blowing vision. But there is one rather serious problem with it. If our universe was created by superior beings in another universe and theirs in turn was created by superior beings in an earlier universe, and so on, then who or what created the very first universe?

One possibility, says Harrison, was that it was God. At first sight this seems a rather weak admission. After all, Harrison came up with the idea of the natural selection of universes specifically to avoid the other explanations of the universe's fine-tuning. One was that there is an infinity of barren universes and the other was that God did it. Harrison, however, sees an important distinction between his idea and the religious view. "In my scheme, God starts things," he says. "Thereafter, however, superior beings in universes take over the creation of further universes."

One other possibility for the origin of the first universe is a variation on the barren-multiverse idea. In the beginning, says Harrison, there might have been a large ensemble of universes, each with its own random variant of the laws of physics. Most of the universes were dead and uninteresting. But, by chance, the conditions were right in at least one for the evolution of life. Harrison calls this the intelligent "mother"

---

* Arthur Eddington, *Space, Time, and Gravitation* (Cambridge: Cambridge University Press, 1920).

universe. "Thereafter, by virtue of the fact they reproduce, intelligent universes come to dominate the ensemble," he says. "In time, the original unintelligent universes become a vanishingly small fraction of the whole."

We are still left with an unanswered question. If a Supreme Being made the first universe, who or what made the Supreme Being? And if it all began with a mostly dead ensemble of universes among which happened to be the intelligent mother universe, how did the initial ensemble come about? "Perhaps the Supreme Being occupied another universe created by an even higher form of intelligence, and perhaps the initial ensemble consisted of botched and bungled creations by a sorcerer's apprentice in another universe," says Harrison.

Here, Harrison is alluding to the words of the philosopher David Hume, who in 1779 wrote: "Numerous universes might have been botched and bungled throughout an eternity, ere this system was struck out; much labor lost, many fruitless trials made, and a slow but continual improvement carried out during infinite ages in the art of world-making."

Could Hume have inadvertently put his finger on how it all began? Who knows? One thing, however, follows automatically from Harrison's vision. If humanity avoids its own destruction and manages to survive into the far future, one day our descendants will have to make a rather important decision: whether or not to become parents.

Humanity is unlikely to be alone in making this decision. For if, as Harrison suggests, our universe was designed by life specifically so that it would give rise to life, then it is likely that other intelligences in other galaxies will sooner or later face a similar dilemma. Which prompts a rather obvious question. Where are the other intelligences? So far, we know of only one example of biology: our own.

The view of most astronomers is that the most likely place to find extraterrestrial life is on Earth-like planets warmed by Sun-like stars. However, a planetary scientist in California thinks that most astronomers could be wrong. According to David Stevenson, the majority of the universe's life may not reside on cozy planets like our own at all. Far from it. If he is right, the place to look for extraterrestrials may be the most hostile environment it is possible to imagine: the super-cold vacuum of interstellar space.

Part Three

Life and the Universe

10

The Worlds between the Stars

*Billions upon billions of habitable planets could be hiding
in the cold, dark abyss of interstellar space*

It was not until Al Worden found himself on the far side of the
Moon and saw all those stars—God, there were so many stars—
that he was sure there was more to the universe than he had ever
imagined. This light show was enough to convince him that it had
to be all or nothing: either there was no life out there or the cosmos was teeming with it.
　　　　　　　　　　　　　　—Andrew Chaikin, *A Man on the Moon*

Do there exist many worlds, or is there but a single world? This is
one of the most noble and exalted questions in the study of Nature.
　　　　　　　　　　　　　　—St. Albertus Magnus

Running into the planet out here was nothing short of a miracle.
The odds against finding it in the vast black chasm between the
stars were literally astronomical. But find it they have. As the starship
spirals in to take a closer look, its crew crowds the observation deck,
straining every sense. The planet, black against black, is barely visible
until a blinding light explodes, bathing the sunless world in more illumination than it has received in a billion years. The blistering fireball
of a landing craft streaks down through the dense atmosphere. In a few
minutes, when radio silence is broken, its pilot will no doubt report a
deep-frozen world where nothing, not even a breath of wind, has stirred
for a million or more centuries. Imagine everyone's shock, then, when
the radio link bursts into life and what the pilot actually reports—no,
yells out at top voice—is: "My God, you've got to see it—it's teeming
with life down here!"

Could life really exist on a world drifting in the space between the stars? A world without a sun to warm it? A world whose stygian gloom is lifted only by the stutter of lightning and the cherry-red glow of lava spilling from volcanoes? The prospects for biology in such a godforsaken location would seem pretty bleak. Appearances, however, may be deceptive. According to David Stevenson of the California Institute of Technology in Pasadena, worlds wandering between the stars may turn out to be the most common sites for life in the whole universe.

The idea that planets might exist in the cold and dark of interstellar space is a novel one. Although astronomers now know of more than sixty planets — nine in our own solar system and fifty or so recently discovered orbiting nearby stars — every last one of them circles a sun like a moth around a campfire. Nobody has yet discovered an "interstellar planet," an orphan world roaming through space without a stellar parent.*

Absence of evidence, however, is not necessarily evidence of absence. Planets, by their very nature, are small, insignificant things — at least compared with stars. They shine solely by the light they reflect from external sources.† An interstellar planet, with no sun to illuminate it, would be effectively invisible. Even if such worlds were common, we would never have found them.

So, why believe in them? The answer is that they pop up naturally in computer simulations of the birth of planetary systems like our own solar system. Such simulations mimic the processes that are thought to go on in an embryonic, or "protoplanetary," nebula of gas and dust swirling around a newborn sun. As time passes, dust particles in the nebula stick together to make bigger particles, which stick together to make yet bigger particles, and so on. This process leads inevitably to the formation of a handful of planet-sized bodies. Some are giants like

---

    *   There has, however, been a recent claim of free-floating planets in the star cluster called Sigma Orionis (M. R. Zapatero-Osorio et al., "Discovery of Young, Isolated Planetary Mass Objects in σ Orionis Cluster," *Science*, 6 October 2000, 103–7). Nobody yet knows whether the objects are real planets or "brown dwarfs," a species of failed star. However, theorists believe that encounters between pairs of stars in clusters may eject some of each star's planets. Certainly, observations of dense "globular clusters," where stellar encounters are common, reveal no planets orbiting stars, and one explanation is that a large proportion have been ejected into interstellar space.

    †   Strictly speaking, this is not true of "giant" planets. In our own solar system, Jupiter gives out about twice as much energy as it receives from the Sun, and Saturn radiates marginally more than it receives from the Sun. This energy is a relic of the formation of the two giant planets. It is given out as infrared, a form of light invisible to the naked eye.

Jupiter while others are more like the size of Earth. The surprise is that the simulations often show a protoplanetary nebula spawning not one body of Earth-mass but more like ten.

This is in clear contradiction to what we see in our solar system today. Instead of ten or so Earth-mass planets, there is only one—two, if you count Venus, which is only a little less massive than Earth. What, then, happened to Earth's brothers and sisters?

The clue comes from where in the protoplanetary nebula Earth-mass planets form. According to the computer simulations, the majority are born in the vicinity of the giant planets. Earth-mass bodies are little more than the "builders' rubble" left over from the construction of Jupiter-like planets.

An Earth-mass body that orbits a sun in the neighborhood of an embryonic giant planet is in grave danger. Sooner or later, it will stray too close to its monstrous neighbor. When this happens, the gravity of the giant planet will catapult it clean out of the planetary system into the depths of interstellar space. This kind of thing must have happened in our own solar system shortly after it was born four and a half billion years ago. "The implication is that Earth might have ten similar-sized brothers and sisters, now irrevocably lost in deep space," says Stevenson.

Stevenson points out, however, that some of these objects could have collided with the surviving giant planets or been swallowed up by the Sun. "My main point is not to emphasize the exact numbers of lost planets but the real possibility that they exist," he says.

According to recent surveys, about one in ten nearby stars has a planetary system. If, for the sake of argument, the formation of each planetary system involved the ejection into interstellar space of about ten Earth-mass planets, it follows that there could be as many interstellar planets in the universe as there are stars. "That means about two hundred thousand million in our Milky Way alone," says Stevenson. "Most Earth-size planets may exist in total darkness."

Energy without Sunlight

The idea of Earth-mass bodies flung into the void between stars during the formation of a planetary system is not a particularly controversial one. What is controversial, however, is the idea that such planets could provide an abode for life. It's fair to say that, before Stevenson came up with the suggestion, nobody dreamed of such a possibility. And with good reason. After all, life requires a sun to supply it with energy, doesn't it?

Not necessarily.

Although it is certainly true that most of the living things around us depend on the energy of sunlight—plants exploit it directly through photosynthesis and animals indirectly by eating plants and other animals—some organisms can and do survive perfectly well without it. In the past few decades, for instance, biologists have been stunned to discover life in some of the harshest of terrestrial environments. For instance, they have discovered colonies of exotic creatures, including meter-long tubeworms, living in superheated water near volcanic vents on the deep sea floor. They have also found bacteria eking out an existence entombed in solid rock many kilometers beneath our feet.

Contrary to all expectations, life can survive and even flourish in complete and utter darkness. Instead of the energy of sunlight, it feeds on something else, most often the energy locked up in sulfur-based chemicals. In the case of the creatures around subsea volcanic vents, this energy is extracted from minerals dissolved in water, and in the case of bacteria deep down in Earth's crust it comes from the solid rock itself.

There is now a strong suspicion among biologists that life on Earth may actually have gotten started at volcanic vents on the sea floor. Certainly, when Earth was young and still under machine-gun bombardment from the leftover debris of planet formation, the seabed would have been a relatively safe haven. The same can be said for the rocks beneath our feet. Bacteria may simply have crept deeper and deeper into Earth in search of a quiet life. Some people have even speculated that there might exist a "deep biosphere," extending way down into Earth and limited only by the extreme temperature found at great depths. Conceivably, the deep biosphere could dwarf the familiar biosphere of surface, oceans, and atmosphere.

The discovery of creatures surviving in the absence of sunlight has not only changed biologists' view of life on Earth, it has caused them to look with fresh eyes at the other planets of our solar system. Several decades ago, the accepted view was that Earth was the only place benign enough for life. Now there is a growing optimism that simple microorganisms might survive in relative warmth in the rocks beneath the frozen surface of Mars. And who knows what might be swimming in the biggest ocean in the solar system—the ice-covered sea of Jupiter's giant moon, Europa?

Looking beyond the borders of our solar system, it is clear that mere lack of sunlight does not in itself rule out the possibility of life on a world wandering alone between the stars. Living things may simply exploit some alternative source of energy. But energy, it turns out, is not the only prerequisite for life. It needs warmth as well.

Warmth without Sunlight

Warmth is essential to keep water from freezing solid. Water is the medium in which the chemical reactions of biology go on. And, as far as we know, it is essential for life. Even bacteria deep in Earth's crust depend on tiny amounts of water trapped in fissures in the rocks.

On Earth, the warmth to keep water from freezing comes mostly from the heat of the Sun. The expectation is therefore that a world without a sun would be far too cold for water and, consequently, life. "However, this is not necessarily the case," says Stevenson. "There are important sources of warmth other than sunlight."

Take Jupiter's giant moon, Europa, for example. Because of its location in the outer solar system, the sun in its sky is twenty-five times fainter and weaker than the one that warms and illuminates Earth. By rights, Europa's ocean should be frozen solid. However, the moon has another source of warmth. As it circles Jupiter, the giant planet's gravity alternately stretches and squeezes the moon and this generates enough heat to keep the bulk of its ocean a liquid.* Closer to home, the interior of Earth is still molten four and a half billion years after the fires of the planet's violent birth have gone out. The reason is radioactivity. Earth's rocks contain radioactive atoms—principally, uranium, thorium, and potassium. As these gradually disintegrate, or "decay," they release large quantities of heat energy, which turn the planet's interior to the consistency of molasses.

Like Earth, an interstellar planet—made of rock or a mixture of rock and ice—would be warmed by the decay of radioactive atoms deep in its core. The energy available from radioactive rocks is a mere

---

* At least this is what scientists suspect. Nobody has actually seen the ocean beneath the ice of Europa. However, NASA's Galileo probe has photographed jigsawlike patterns on the ice that look for all the world like the patterns in free-floating sea ice on Earth. Europa also appears to have a magnetic field, which can best be explained by electrically conducting salt circulating in a global ocean beneath the ice.

ten-thousandth that is available from sunlight on Earth. However, according to Stevenson, this is enough to make a big difference to the planet. The problem, it turns out, is not generating heat but hanging on to it.

Clinging to Warmth

In the space between the stars it is mind-cringingly cold—typically -260 degrees Celsius. Because of the huge difference in temperature between the molten interior of an interstellar planet and its hostile surroundings, heat will gush out into space. The only way the surface could possibly stay warm enough for liquid water to exist would be if it were wrapped in some kind of blanket that drastically impeded the rate at which heat escaped.

Earth possesses just such a blanket. Although the planet's surface is warmed by sunlight, by rights much of this warmth should leak away into space. What stops it, however, is moisture, or water vapor, hanging in the atmosphere. Water vapor is an extremely effective "greenhouse" gas that absorbs and traps outgoing heat. Without its presence in our atmosphere, the average temperature of Earth's surface would be about -40 degrees Celsius.*

Water vapor may be perfectly adequate for keeping Earth from freezing. However, the requirements of an interstellar planet are of a different order. To hold onto its meager heat it would need a greenhouse blanket hugely more effective than that of Earth. Remarkably, however, such a blanket is possible. "In fact, it is positively guaranteed," says Stevenson. "What comes to the rescue is molecular hydrogen, the most common constituent of the nebula from which planetary systems form."

Molecular hydrogen is the lightest of all gases. When Earth was newly formed, it clung to the planet. However, the Sun warmed the gas so that it soon floated to the top of the atmosphere and wafted away into space. According to Stevenson, the fate of the mantle of molecular hydrogen swaddling an interstellar planet would be rather different. In the deep freeze of interstellar space there is no warmth to drive the gas

---

*   The most famous greenhouse gas is carbon dioxide, which is liberated in the burning of fossil fuels such as coal and oil, and is implicated in "global warming." Water vapor, however, is a far more potent greenhouse gas than carbon dioxide.

route. Initially, his interest was in something a little more mundane—dust. Or, more precisely, the dust between the stars.

Ever since the 1920s, astronomers have known that the stars are not so bright as they ought to be. In every direction that telescopes are pointed, starlight is dimmed by dust, hanging like a shimmering veil across the night sky. The grains of interstellar dust were known to be small—typically a thousandth of a millimeter across, which is about the size of a typical bacterium.* The big question was: what were those grains made of?

One early suggestion was iron, because some of the meteorites that fall to Earth are solid iron. However, this was later ruled out when it was realized that there was simply too little iron in interstellar space to account for the enormous amount of dust. The other possibility was that the dust grains were some combination of the commonest atoms in interstellar space—hydrogen, carbon, oxygen, and nitrogen. One plausible candidate was water ice (H_2O), but in the early 1960s this too was ruled out. Not only was it difficult to see how ice could accumulate in an ultrararefied cloud of interstellar gas, but ice strongly absorbs infrared light, a type of light invisible to the naked eye, at a wavelength of 3.1 micrometers.† No such absorption was discovered when the first infrared telescopes were pointed at the heavens.

Enter Wickramasinghe. In 1960, he arrived in Britain from his native Sri Lanka to work with Fred Hoyle, arguably the most influential British astronomer of the postwar era. For his Ph.D. at Cambridge, Wickramasinghe considered the possibility that the mysterious interstellar dust grains were made of graphite, a pure form of carbon used as the "lead" in pencils. Things looked promising. Wickramasinghe's calculations based on new experimental data for graphite in the lab showed that the material affected light in just the right way to explain the observed dimming of starlight. So convincing was the case for

---

* Astronomers were able to guess the size of the dust grains because of a simple argument. Grains about a thousandth of a millimeter across are the best at "scattering," or redirecting, starlight. Consequently, the observed dimming of starlight could be explained with a relatively small quantity of dust. If the grains were much smaller or bigger, they would be far less efficient at scattering starlight and more dust would be needed than could possibly exist in interstellar space.

† Light is actually a wave motion like a water wave. It is characterized by a "wavelength," related to the distance between successive crests or troughs of the wave. Infrared light is distinguished from visible light by having a slightly longer wavelength.

graphite that for several years after Wickramasinghe completed his Ph.D. in 1962, the general opinion among astronomers was that the problem had been solved.

A spanner was thrown in the works, however, in the early 1970s when astronomers discovered that interstellar dust grains absorb infrared light at 3.4 and 10 micrometers. Graphite does not absorb starlight at such wavelengths. Wickramasinghe was forced back to the drawing board.

Carbon, in addition to making graphite, can make an awful lot of other substances as well. This is because it has a unique ability to combine with itself and other atoms. So good is it at linking up to make complicated molecules that chemists have divided their subject into "organic" chemistry, which deals with all the carbon-based molecules, and "inorganic" chemistry, which deals with the rest. Wickramasinghe decided to test some organic compounds. He settled on "organic polymers": long, daisy-chain-like molecules in which a carbon-based group of atoms is repeated over and over again.

By now, Wickramasinghe had moved to the University of Wales College, Cardiff. The results he obtained there in 1974 were encouraging indeed. The first organic polymer tested turned out to be the best mimic yet of the mysterious interstellar grains.

At the time, Hoyle was in Pasadena visiting colleagues at the California Institute of Technology. Wickramasinghe wrote to him with the good news. He ended his letter with an extraordinary question: "Should we be thinking about an origin of life in interstellar space?"

Wickramasinghe's reasoning was simple. His experiments strongly indicated that the interstellar grains were made of complex organic molecules, and complex organic molecules are the basic building blocks of life. Had Wickramasinghe been writing to anyone other than Hoyle, he might have been a little more cautious. However, Hoyle was the one person in the world who was likely to take Wickramasinghe's outlandish suggestion seriously. For, in addition to being the most prominent astronomer of his generation, he had also dabbled in the writing of science fiction. In his most celebrated novel, *The Black Cloud*, he had imagined the arrival in the solar system of an intelligent cloud of interstellar hydrogen—life from the very depths of space.

Wickramasinghe's letter went unanswered for several weeks. Finally, a reply came from southern California. In it, Hoyle marshaled every argument he could think of against the idea of life coming from space.

Despite his reservations, however, Hoyle encouraged Wickramasinghe to pursue the idea. And, in 1977 and 1978, the two astronomers submitted a series of papers to the international science journal *Nature*. In them, they explored the idea that the first steps from nonlife to life had been taken not on Earth but in interstellar space. The details were as yet hazy. But what the astronomers were advocating was a "prebiology," an extraterrestrial precursor to conventional biology without which life on Earth could never have got going.

Perhaps unsurprisingly, given their explosive content, the papers in *Nature* were met with stunned silence. "As soon as we dared to mention the possibility of biology on a galactic scale, the scientific community closed ranks to stop further publications," says Wickramasinghe.

It was shortly after this that Wickramasinghe had a brainstorm. He could spend a great deal of time testing the light-absorbing properties of one organic polymer after another in the hope of finding an even better mimic of interstellar dust. Or he could take a shortcut: use bacteria. Bacteria, after all, were tiny biochemical factories, chock-a-block full of different organic polymers. By shining light through the dried-out hollow husks of bacteria, he could test lots of organic polymers simultaneously.

In 1978, Wickramasinghe and his Iraqi research student Shirwan Al-Mufti carried out the experiment. The result was spectacular. The effect on the light was the closest yet to the effect of interstellar grains. As a bonus, the experiment suggested a prediction. According to the laboratory results, dried bacteria also absorbed infrared light in a characteristic way between two and four micrometers. Here was their unmistakable fingerprint.

It was an observational test by which the idea would either live or die. What was needed was a bright beacon of infrared light out in space. If the beacon were far enough away, its light on its way to Earth would stream through great swathes of interstellar dust. Hoyle knew the perfect beacon: it was called GC-IRS7 and it was a powerful source of infrared light, shining out from the very heart of our Milky Way galaxy. All the astronomers needed to do was to point a big telescope at it and look for the telltale infrared absorption between two and four micrometers.

The center of the Milky Way is visible only from the southern hemisphere. Fortunately, Wickramasinghe's brother, Dayal, was an astronomer in Australia. Dayal applied for observing time on the giant Anglo-Australian Telescope in New South Wales, but nobody took the project seriously and the proposal was turned down. However, Dayal happened to have been awarded some observing time on the AAT for another project entirely. In between his planned observations, he managed to squeeze in an observation of GC-IRS7, obtaining an infrared spectrum, which he faxed to his brother in the United Kingdom.*

When Wickramasinghe and Hoyle saw the spectrum emerging from their fax machine, they were stunned. Staring out at them was the unmistakable fingerprint Wickramasinghe had seen in the laboratory—the fingerprint of extraterrestrial biology.

It was always possible that the grains floating in space were something other than bacteria. Wickramasinghe and Hoyle conceded this possibility. However, if the grains were something else, they argued, that something else not only had to be the size of a bacterium but it also had to absorb starlight just as bacteria did. The simplest option, they maintained, was to accept the bacterial option, no matter how outrageous it appeared.

The irony was that for years astronomers had compared the size of interstellar grains to the size of bacteria. It was simply a convenient size comparison. They did not read any more into it. Nor at the outset of their work did Wickramasinghe and Hoyle. Now, however, they were saying that the coincidence was no coincidence at all. The reason that interstellar grains were the size of bacteria was that they were bacteria.

Interstellar space was a vast graveyard for the microorganisms.

The two astronomers announced their discovery at scientific meetings and conferences, convinced that everyone would be bowled over by the weight of the evidence. "Everywhere we went, however, we got an uncomfortable reception," he says. "People were stunned and speechless."

In retrospect, Wickramasinghe admits that they were rather naïve. It was going to take a lot more evidence to convince the skeptics. Wickramasinghe and Hoyle therefore set out to flesh out their extraordinary idea.

It was immediately obvious to them that, if interstellar space was a graveyard for bacteria, there had to be a stupendous quantity of bacteria out there. The interstellar dust grains were known to account for at least a third of all the carbon in interstellar space. That made the total mass of the grains in our galaxy about ten million times the mass of the Sun. Wickramasinghe and Hoyle were therefore claiming that some-

---

* A spectrum of a light source is simply a graph of how much light is produced at each wavelength.

thing like a ten-thousandth of the mass of the Milky Way was in the form of dead bacteria.

Generating such a tremendous mass of bacteria poses a huge problem in practice. However, it poses no problem whatsoever in principle. Bacteria are capable of a quite phenomenal rate of reproduction. Typically, one can split into two in two to three hours. At this rate of doubling, after four days a single bacterium could produce a million million offspring—enough to fill the volume of a sugar cube. After another four days, its descendants could fill a village pond. After four days more, the volume of the Pacific Ocean. In less than two weeks, a single bacterium could have converted itself into a mass of bacteria equivalent to the mass of the entire Milky Way.

Bacteria can proliferate at such a runaway rate only if they have access to abundant nutrients. After all, the building of new bacteria requires a supply of chemical building blocks as surely as the building of new houses requires a supply of bricks and mortar. The big question was, where in the galaxy is there such a supply? Where, in other words, do the untold legions of interstellar bacteria come from? Wickramasinghe and Hoyle could think of only one plausible place: comets.

The Comet Connection

Comets are dusty snowballs the size of small mountains. They are the builders' rubble left over from the formation of the Sun and planets four and a half billion years ago. Our solar system is believed to contain about a hundred billion of them. They orbit the Sun silently, invisibly, in a vast cloud beyond the outermost planet, Pluto. This Oort Cloud is so huge that it extends a fair fraction of the way to the nearest star.

Most, if not all, of the two hundred billion other stars in our galaxy are thought to have their own comet clouds. Assuming that each cloud has a mass similar to that of the Oort Cloud—roughly that of a medium-sized planet such as Uranus or Neptune—it is possible to estimate the total mass of comets in our Milky Way. The answer is about ten million times the mass of the Sun. Wickramasinghe points out this is almost exactly the same as the mass tied up in dead bacteria in interstellar space.

If nothing else, this mass coincidence shows that comets possess sufficient material to supply interstellar space with bacteria. However, it hardly points the finger at comets as the source of such material. For proof of a bacteria-comet connection, Wickramasinghe and Hoyle point to a number of other pieces of evidence. One is that comets have a ready-made mechanism for ejecting bacteria-sized particles into interstellar space.

The huge majority of comets spend their entire lives orbiting far from the warmth of the Sun in the deep freeze of space. Occasionally, however, one is nudged, perhaps by the gravity of a passing star or maybe by a collision with another comet. Whatever the details, the outcome is that the comet begins to fall sunward. Faster and faster, it falls until it hurtles past the planets and enters the inner solar system. There, the warmth of the Sun begins to boil off the ice and dust on its surface. Before it swings around the Sun, the comet, once dark and inert and too faint to be seen by even the most powerful telescope, develops a multimillion-mile-long tail of expelled debris that reflects sunlight and glows spectacularly.

Typically, such a comet ejects a million tons of debris a day into interplanetary space. However, the material does not necessarily stay there. The reason is sunlight. Sunlight is like a perpetual wind blowing out past the planets. Although it is too weak to affect large bodies, it sweeps tiny bodies clean out of the solar system. How tiny? Well, bacteria-size particles are certainly tiny enough. Consequently, any that are ejected by a comet are promptly blown into the depths of interstellar space.

In the beginning, Wickramasinghe and Hoyle had no firm evidence that comets were ejecting bacteria-like particles. However, in 1986, a flotilla of unmanned space probes flew past the icy "nucleus" of Halley's Comet on its once-every-seventy-six-years visit to the inner solar system. Instruments on board the space probes revealed that the particles boiling off the nucleus were not only the same size as interstellar dust grains but they also absorbed light in the same way.

In the mid-1970s, however, all Wickramasinghe and Hoyle knew for sure was that there existed a well-understood mechanism for getting bacteria-sized particles from comets out into the depths of interstellar space. This was all they needed, however. If, as the two astronomers believed, bacteria really did infest comets, it was unavoidable that over the billions of years the galaxy had existed, huge numbers of bacteria had been driven out into the vacuum between the stars.

But why would bacteria infest comets? How would they find their way into these icy bodies in the first place? The answer, according to Wickramasinghe and Hoyle, is to be found in the process of star birth.

In the accepted picture, a cloud of interstellar gas and dust begins shrinking under its own gravity. Possibly this shrinkage is started by the blast wave from a nearby exploding star, or "supernova." Whatever the trigger, however, the cloud shrinks faster and faster, in the process becoming denser and hotter. In the latter stages of cloud collapse, large bodies begin to condense out—comets, planets, and a central star.

Here, then, is how bacteria get into comets. If, as Wickramasinghe and Hoyle believe, they are present in interstellar clouds, inevitably they will be incorporated in all the bodies that congeal out of such clouds. Any bacterium assimilated into a star or planet, however, will be quickly vaporized by the intense heat associated with its birth. Only in the relatively benign environment of comets will bacteria remain intact.

But what is the use of bacteria remaining intact if they are all dead to start with? After all, in Wickramasinghe and Hoyle's view, interstellar space is a graveyard of microorganisms. Clearly, the only way bacteria can multiply is if a small number are viable so that once inside a comet they can return Lazarus-like from the grave.

This is a tall order. Few environments are quite so harsh and unforgiving as interstellar space. Bacteria must survive empty vacuums, mind-numbing cold, and intense cosmic radiation for millions of years, perhaps even billions of years.* However, Wickramasinghe and Hoyle point to the remarkable abilities of terrestrial bacteria known as "extremophiles." These are able to survive dehydration, temperatures of more than a hundred degrees below freezing, and intense radiation. One species, known as Deinococcus radiodurans, can survive in the cores of nuclear reactors, repairing even extensive damage to its DNA so that it can replicate normally. In principle, bacteria could survive the much lower radiation background of interstellar space for billions of years.

Because of the phenomenal ability of bacteria to multiply, only a tiny fraction of interstellar bacteria needs to be viable. According to Wickramasinghe and Hoyle, even a vanishingly small concentration— say, one in a hundred million million—would be enough to convert a sizeable fraction of the mass of a comet into microorganisms within a matter of days. The proviso is that there exist two things: liquid water and a large supply of organic molecules.

---

* Cosmic rays are high-speed atomic nuclei, mostly protons. Low-energy ones come from the Sun, high-energy ones probably from supernovae. The origin of ultra-high-energy cosmic rays, particles millions of times more energetic than anything we can currently produce on Earth, is one of the great unsolved puzzles of astronomy.

In the early 1970s, it was a leap of faith to suggest that organic molecules were present in comets. Such bodies were known to contain ices of simple compounds such as water, methane, and ammonia, but that was about all. However, by the 1980s, radio telescopes had detected scores of ever more complex organic molecules in interstellar space. And there was a strong suspicion that these were merely the tip of the iceberg and that out in space there existed more-complex molecules, perhaps even the amino-acid building blocks of DNA.*

The relevance of all this to comets was that the complex organic molecules detected in space would undoubtedly be incorporated into comets when these bodies congealed out of interstellar gas clouds. It was rather indirect evidence. However, direct evidence of the presence of organic molecules in comets came in 1986 when Halley's Comet plunged into the inner solar system. The European space probe, Giotto, and the Russian Vega probes flew close to the peanut-shaped nucleus of the comet. They found that its surface was as black as coal, as would be expected if it were coated in organic material. Wickramasinghe and Hoyle had predicted this. Everyone else had expected a gleaming white nucleus, in keeping with the picture of a comet as a "dirty snowball."

So much for a supply of organic molecules in comets. What about a supply of liquid water? Bacteria, in common with human beings and all other living things, are composed largely of water, and water is the universal "solvent" in which the chemicals of life can float about freely and react together. Without it, life—or at least the DNA-based life with which we are familiar—is impossible.

Here Wickramasinghe and Hoyle's ideas would seem to founder rather badly. After all, far out in the Oort Cloud, with the Sun little more than an unusually brilliant star in the sky, the typical temperature is more than two hundred degrees below zero. Comets are as frozen as frozen can be. They patently do not contain liquid water.

However, Wickramasinghe and Hoyle claim that the conditions in the Oort Cloud have not always been so harsh. Long ago, when the solar system was young, they say, comets were warm. The source of their warmth was "radioactivity"—specifically, the radioactivity of a particular type of aluminum known as aluminum-26.

---

* The simple amino acid glycine was detected in an interstellar cloud at the center of our galaxy in 1994. Though glycine is not one of the amino acids used by living organisms, its discovery nevertheless raises hopes that amino acids more relevant to life exist in space.

Aluminum-26 is forged in the fireballs of supernova explosions. Supernovae are thought to be intimately connected with star birth, either because their blast waves actually trigger the collapse of gas clouds to form stars or because some of the stars in stellar nurseries—the most massive ones—race through their lives and go supernova before some of the others have had a chance to be born. Either way, supernovae inject aluminum-26 into the raw material from which stars, planets, and comets congeal. We know that this actually happened to the cloud that spawned the Sun and planets. The evidence comes from the "decay products" of aluminum-26, which geologists find in meteorites, the rocky debris left over from the birth of the solar system.

The upshot of all this is that, when comets are born, they contain aluminum-26. However, aluminum-26 is radioactively unstable. As time passes, more and more of its atoms disintegrate. This process of "radioactive decay" liberates large quantities of "nuclear energy"—more than enough, say Wickramasinghe and Hoyle, not only to melt the interiors of comets but to keep them in a melted state for several million years.

According to the two astronomers, a newborn comet would have had a solid exterior and a liquid center, just like a giant chocolate liqueur. Beneath its frozen surface—perhaps a few hundred meters, maybe only a few tens of meters, down—would exist a giant underground swimming pool—a swimming pool in which the conditions would be ideal for a small number of interstellar bacteria to embark on a runaway orgy of reproduction.

Summertime in the Oort Cloud would have been brief, however. Aluminum-26 decays with a half-life of 740,000 years, which means that after 740,000 years half would have remained undecayed, after another 740,000 years a quarter, and so on. With the precious heat source fading away, the liquid interiors of comets would have been plunged into winter within a few million years of their formation.

The freezing of comets might be expected to kill the bacteria, which had proliferated so wildly in balmier times. After all, water expands when it turns to ice and this can kill bacteria by bursting their cell membranes, the bags that hold in their chemicals. However, Wickramasinghe says this happens only if freezing is sudden. Because the comets' heat source would die away only gradually, freezing would occur far more slowly. There would be time for water to "diffuse" out of bacteria so that when they finally froze there would be no ice left inside to rupture their membranes. Wickramasinghe and Hoyle therefore claim that some bacteria would remain viable.

Clearly, this is essential to Wickramasinghe and Hoyle's idea. After all, for comets to be seeded by interstellar bacteria, some of those interstellar bacteria must be capable of revival. And for some interstellar bacteria to be viable, some of the bacteria supplied by comets to interstellar space must also be capable of revival.

But if the bacteria inside comets remain viable even after the comets freeze, there is another spin-off: a possible explanation for one of the greatest mysteries of all. The mystery is the origin of life on Earth.

Life on Earth

The origin of terrestrial life poses a deep mystery because it appears to have got started so quickly. For hundreds of millions of years after Earth was formed, the planet was in a fiery, semimolten state far too hostile for life. It wasn't until about 3.85 billion years ago that it had cooled down sufficiently for pools of liquid water to exist on the surface without boiling away instantly. And it was at this moment—the earliest moment that it was possible—that life seems to have made its first appearance on Earth.*

The clear message from this is that life must be easy to start. And herein lies a puzzle. Scientists have gone to enormous pains to reproduce the conditions they believe existed on the ancient Earth. In numerous laboratory experiments since the 1950s, they have recreated the "primeval soup" of chemicals and energized it with electrical discharges to mimic the effect of ancient lightning. But, try as they might, they have been unable to create life from nonlife.

This poses a dilemma. How is it that we find it so difficult to start life when it burst forth with such apparent ease on the primitive Earth? One possibility is that we have somehow overlooked an essential ingredient

---

*   The evidence of this does not come from fossils, because soft-bodied bacteria leave no fossils. It comes instead from a curious anomaly in the rocks of this era. Carbon comes in several different forms, known as "isotopes." The most common is carbon-12, but other forms such as carbon-13 can also exist. Living organisms tend to concentrate carbon-12. So when they die and become incorporated in rocks, those rocks have an unusually high concentration of carbon-12 relative to carbon-13. An elevated level of carbon-12 is indeed found in rocks from 3.85 million years ago. They show the fingerprint of life.

in the primordial chemical mix. After all, we are only guessing what it was like on Earth 3.85 billion years ago. However, Wickramasinghe and Hoyle have a more radical explanation for the failure of laboratory experiments. The reason nobody has been able to create life from non-life, they say, is that it is extremely hard to do.

But if it is so hard to do, how come life got started so quickly on Earth? There is only one way out of the apparent paradox, say Wickramasinghe and Hoyle. The step from nonlife to life must have been taken elsewhere. Life, they maintain, did not originate on Earth. It was seeded from space.

The reason life got going so soon, say Wickramasinghe and Hoyle, is that it arrived on the planet ready-made. From the moment Earth was born four and a half billion years ago, life rained down onto its surface from the heavens. Who knows how many comets came, how many times the microorganisms they carried were extinguished? But, finally, about 3.85 billion years ago, when Earth was cool enough and the conditions were right, the seeds of life fell on fertile ground.

The evidence that comets were striking Earth 3.85 billion years ago can still be seen on another body that was in the firing line: the Moon. The giant Mare basins, or lunar "seas," date back to this time, a period of intense bombardment by comets and asteroids. The comets that hit Earth brought water and organic material. But, claim Wickramasinghe and Hoyle, they also carried a far more important cargo: living things.

If the two astronomers are right, our primitive ancestors came from the stars. We are all extraterrestrials.

The idea that life was seeded from space is by no means a new one. It seems to have originated with Aristarchus of Samos in the third century B.C. In the nineteenth century, however, it was championed by two of the greatest physicists of their day: William Thomson in England and Hermann von Helmholtz in Germany. Thomson and Helmholtz claimed that the seeds of life were spread from star to star. This theory of "panspermia" was little more than idle speculation until the work of the Swedish Nobel Prize–winning chemist Svante Arrhenius at the beginning of the twentieth century.

Arrhenius put part of the idea to the test by seeing whether bacterial spores and plant seeds could survive the conditions thought to exist in interstellar space. He arranged for botanists to expose them to conditions of near vacuum and temperatures as low as -196 degrees Celsius, the boiling point of liquid nitrogen. On reheating, despite their terrible ordeal, they came back to life. Nowadays, we know that bacteria can also survive intense ultraviolet light and cosmic radiation of the kind present in interstellar space. As Arrhenius was the first to recognize, bacteria have quite "unearthly" properties.

This poses a mystery. According to Charles Darwin's theory of evolution by natural selection, the traits possessed by a living organism are the traits that ensure its survival in the environment to which it has been exposed—that is, the environment of Earth. Why should bacteria have survival traits for an environment to which they have never been exposed—interstellar space? The standard answer is that this is an accident. They have not been selected for space-worthiness but their space-worthiness is an incidental by-product of some other advantageous trait. Wickramasinghe and Hoyle do not believe this. The reason bacteria have the traits necessary for survival in space, they say, is that they originated in space.

"I believe a leaf of grass is no less than the journeywork of the stars," wrote Walt Whitman. Perhaps he was more right than he could ever have imagined.

One of the consequences of Wickramasinghe and Hoyle's reworking of the old idea of panspermia is really quite remarkable. If bacteria were raining down from the skies at the dawn of time, then it follows that they must be raining down on Earth today. Comets, after all, are continuing to plunge into the inner solar system from the Oort Cloud at a rate of dozens a year. Earth may not have been hit by a comet in recent years—something for which we can be quite thankful—but an actual impact may not be necessary for cometary material to fall onto the planet. There is another way, which may also have had a role in the initial seeding of Earth.

Recall that dust is driven from comets by the heat of the Sun. Some of it is blown by the pressure of sunlight into interstellar space. But some of it hangs around, at least temporarily, in the inner solar system. Earth, as it orbits the Sun, cannot avoid plowing through this material. The planet sweeps up hundreds of tons of interplanetary dust every day. If much of this dust is bacterial, as Wickramasinghe and Hoyle believe, then at this very moment microorganisms are wafting down through the atmosphere. The two astronomers' version of panspermia, far from being a process remote from us in both space and time, is operating on our very doorstep today.

What consequences might this have? Wickramasinghe and Hoyle can think of at least one. It might spread bacterial diseases. Some of the incoming bacteria could be infectious agents. This could explain puzzling instances when a disease breaks out at many places across the globe simultaneously, something that they say is difficult or impossible to explain even in an age of rapid air transport. "It is easy to explain, however, if pathogens are raining down everywhere from space," says Wickramasinghe.

This idea that some diseases come from space is very controversial, to say the least. And no wonder. Not content to invade the turf of biologists, Wickramasinghe and Hoyle have dared to trespass on the territory of doctors and epidemiologists as well.

If interstellar bacteria are indeed raining down from the sky, then there is yet another consequence. All bodies in the solar system, not just Earth, must be subjected to the same biological rain. Currently, there is intense interest in whether primitive life exists on Mars or in the giant ocean beneath the ice of Jupiter's moon, Europa. If Wickramasinghe and Hoyle are right about their bugs-from-space idea, the "exobiologists" will not be disappointed. "Wherever in the solar system the conditions for life exist, we predict life will be found," says Wickramasinghe. "What's more, all of it will be DNA-based, just like ours."

In fact, it is possible to speculate even further. It follows from Wickramasinghe and Hoyle's idea that wherever in the galaxy there are environments suitable for life, there will be life. And organisms from one end of the Milky Way to the other will be fundamentally related—or at least the microorganisms will be.

The Great Cosmic Cycle of Life

In Wickramasinghe and Hoyle's picture, however, the seeding of planets is merely a spin-off of a far greater cosmic cycle—the cycle in which bacteria pass from comets into interstellar space, then back to comets again. To recap, this is how it works.

Stars, together with planets and comets, congeal out of clouds of interstellar gas and dust. The comets incorporate bacteria from the interstellar clouds, a few of which are viable. These thrive and multiply inside the comets. When comets fall toward a sun and the heat boils off their surface ices, bacteria are liberated. Some find their way to the surfaces of planets. But the majority are driven by the pressure of light into interstellar space. New stars, together with new planets and new comets, congeal out of clouds of interstellar gas and dust. And so the cycle goes around again. "We estimate that the whole cycle takes about three billion years," says Wickramasinghe. "In other words, it takes three billion years for a bacterium ejected from a comet to become incorporated in another comet when a new planetary system is born."

The beauty of the whole idea is that life need arise only once in the galaxy. When this happens, there is a ready-made mechanism in existence to amplify it and spread it to every last planetary system. In this way, the apparently irreconcilable is reconciled. Life can simultaneously be extremely hard to get started and ubiquitous.

Those who believe that life arose in isolation on Earth must accept that the step from nonlife to life must be taken over and over again at every location in the galaxy. If that step is a hard one, and requires all sorts of accidents to occur, then those same accidents must happen time and time again. If they do not, Earth is likely to be the only inhabited planet.

The big question is clearly: how did the cosmic cycle of life get started? Where did the first bacterium come from? Here Wickramasinghe and Hoyle freely admit they do not yet know the answer. However, this does not trouble Hoyle. Unlike the majority of astronomers, who firmly believe that the universe was born in a Big Bang twelve to fourteen billion years ago, he believes it has always been in existence.[*] In an eternal universe, with an infinity of time to play with, it matters not at all how unlikely is the emergence of life. It will emerge sooner or later with 100 percent certainty. And for Hoyle and Wickramasinghe's scheme to work, it has to emerge only once.

If the two astronomers are right, life is not a mere planetary phenomenon, as most scientists believe. It is a cosmic phenomenon. Far from being an inconsequential spin-off of the laws of nature, it is a central player in the universe. This raises the biggest puzzle of all. Why is the universe so set up for the spread of life?

Now there indeed is a question.

---

[*] The irony here is that it was Hoyle who, in a BBC radio broadcast in 1950, coined the term "Big Bang."

So much for primitive life in the universe; what about the prospects for advanced life? One rather depressing possibility is that bacterial life may be extremely common while intelligent life is extremely rare. On Earth, life spent many billions of years at the primitive, single-cell stage. This may indicate that the step up to more complex, multicellular organisms is a difficult one and that Earth may be one of the only places where this has so far happened. This is not a view that Hoyle and Wickramasinghe subscribe to, because they are convinced that the universe, for some unknown reason, is geared up for life.

If, however, there are other intelligences abroad in the universe, then the pressing question is: where are they? The obvious place to look—so obvious that it seems ridiculous even to mention it—is in the sky. However, an astronomer in the Ukraine thinks we might profitably start our search a little closer to home. His name is Alexey Arkhipov, and the extraordinary place he advocates looking is beneath our feet. According to Arkhipov, the aliens may not have come here yet but they may have sent their garbage ahead of them.

12

Alien Garbage

*The aliens might not have got to Earth yet
but their garbage may have arrived ahead of them*

The object before which the space-suited man was posing was a
vertical slab of jet-black material, about ten feet high and five feet
wide; it reminded Floyd, somewhat ominously, of a giant tomb-
stone. Perfectly sharp-edged and symmetrical. It was so black it
seemed to have swallowed up the light falling on it; there was no
surface detail at all. It was impossible to tell whether it was made
of stone, or metal, or plastic—or some material altogether unknown to man.

—Arthur C. Clarke, *2001: A Space Odyssey*

In the movie *2001: A Space Odyssey*, an alien artifact is dug up on the
Moon. Exposed to the light of the lunar dawn for the first time in
millions of years, it promptly broadcasts a message to the stars: "I've
been found!" Long ago, when its extraterrestrial makers swept through
the solar system, they observed the abundant life on the third planet
from the Sun and recognized its potential. They could not stop with so
many other stars to explore. However, they buried a "sentinel" on the
Moon, a kind of burglar alarm to warn them if one day intelligence
arose on the third planet, sidestepped nuclear annihilation, and ven-
tured out of its cradle into space.

Is it likely that there really are alien artifacts buried beneath the surface of the Moon, or even the surface of Earth? The astonishing answer, says Alexey Arkhipov of the Institute of Radio Astronomy in
Kharkov, is yes. If intelligent life elsewhere in our galaxy has arisen and
made the leap into space, the presence of alien artifacts on our cosmic
doorstep is not only likely but guaranteed.

The claim, at first sight, seems utterly mad. However, it is important to understand what is and isn't being claimed by Arkhipov. He is not claiming that alien artifacts have been left on Earth deliberately, as they were on the Moon in 2001. That would require intention, and who are we to guess the intentions of an extraterrestrial civilization, far in advance of our own? No, the alien artifacts Arkhipov is actually referring to are ones that have fallen to Earth accidentally.

How could artifacts get here accidentally? For an answer, says Arkhipov, we need look no farther than our own cosmic backyard. Human activities in space "pollute" our solar system. Agencies in the space business have become increasingly alarmed by the accumulation in space of dead satellites, discarded rocket casings, and the like. This "space junk" clutters Earth's orbit and is so hazardous to space traffic that it has already led to the postponing of more than one launch of NASA's space shuttle.

But—and this is Arkhipov's point—interplanetary garbage does not stay interplanetary garbage forever. It is inevitable that, over time, some man-made artifacts will actually leave the solar system. Buffeted by the winds of space, they will sail off toward the stars.

Escape from the Solar System

Several distinct processes can eject debris from the solar system, according to Arkhipov. First, there is the pressure of sunlight. Sunlight is actually a wind of tiny particles known as photons, which flood out from the Sun in their countless quadrillions. We are not directly aware of this wind from the Sun because it is so weak.* Nevertheless, it can sweep a body clean out of the solar system—if the body is small enough.

The reason size matters is that to break free of the Sun's powerful gravity, a body must attain an "escape velocity" of more than 600 kilometers per second, or 1.38 million miles per hour. Because the wind from the Sun is so weak, the only bodies that it can accelerate to this speed are very small—no more than a thousandth of a millimeter across. This happens to be the typical size of particles blasted from rocket exhausts. Such particles can be picked up by the photon breeze and blown past the outermost planets and out into the void beyond.

---

* Solar sailing is explored entertainingly in Arthur C. Clarke's short story "The Wind from the Sun," in *The Wind from the Sun* (London: Corgi, 1976).

Bigger bits of debris can also be boosted to the high speed necessary to escape the solar system. According to Arkhipov, this can happen if they collide with each other in space or if they spontaneously explode. In recent years, several planetary space probes are thought to have been destroyed when they either collided with meteorites in space or spontaneously exploded. According to Arkhipov, if such a collision or explosion were to occur in the outer part of the solar system, where the grip of the Sun's gravity is weak, chunks of high-speed debris could be jettisoned into interstellar space.

There is a third means of expelling space debris from the solar system in addition to sunlight pressure and collisions/explosions. This is the close encounter of an artifact with a planet. In such an encounter, known as a "gravitational slingshot," the gravity of the planet actually catapults the artifact into interstellar space.* Computer simulations have shown that, over time, more than a third of the minor planets, or "asteroids," will be ejected from the solar system in this manner.†

Arkhipov's reason for highlighting the mechanisms by which debris can be ejected into interstellar space is to point out that they could just as well act in reverse. In the same way that human activities pollute the solar system with garbage, the activities of any space-faring extraterrestrials will similarly fill their planetary system with space junk. And, just as our technological activities lead to the spread of artifacts into interstellar

---

*   A gravitational slingshot seems impossible. The speed gained by a body such as a spacecraft as it falls through a planet's gravitational field should be exactly the same as the speed it loses as it climbs back out again. There should be no net gain in velocity. This is certainly true from the point of view of the planet. However, it is not true from the point of view of the rest of the solar system. A planet is not a stationary body but instead travels in its orbit around the Sun. Consequently, in any calculations, the velocity of the planet must be tacked onto both the incoming and the outgoing velocities of the spacecraft. In the encounter, the spacecraft "steals" some of the planet's energy. The planet actually gets slowed down in its orbit by a tiny amount while the spacecraft's speed is boosted. This, at least, is the scenario for a spacecraft passing "behind" the planet. If the spacecraft passes in front of the planet, it will be slowed down rather than speeded up.

†   Asteroids are rocky bodies that orbit the Sun between the orbits of Mars and Jupiter. The bodies, of which there are a huge number, range in size from a thousand kilometers across in the case of Ceres down to mere fist-sized stones. Once thought to be the remains of an exploded planet, the asteroids are now believed to be the debris of a planet that never formed because of the disruptive effect of Jupiter's gravity.

space, so too will theirs. The consequences of this, says Arkhipov, are obvious. "If alien artifacts are really floating between the stars," he says, "some of them will inevitably fall to Earth."

Arkhipov's reasoning is based on the absolute inevitability of accidents in space. His only assumption is that our galaxy does indeed contain other space-faring societies. If it doesn't, then his argument is void. But then so too is the argument for SETI, the much-publicized search for extraterrestrial intelligence, which currently involves radio and optical astronomers scanning the heavens in the hope of picking up an intelligent message from the stars. "For Christopher Columbus, the evidence of new lands was strange debris which had floated across the ocean," says Arkhipov. "In the same way, debris which has floated across the ocean of space could provide us with the unmistakable evidence of new planets and new life."

Assuming that there are other space-faring societies in the Milky Way, Arkhipov then poses the following extraordinary question: "During its four-and-a-half-billion-year history, how many alien artifacts could have fallen to Earth?"

How Many Alien Artifacts?

The answer depends on a number of factors. Say each alien civilization in our galaxy has at its disposal the same amount of material as exists in our own asteroid belt. This amounts to roughly 1.8 trillion trillion grams. Now assume that each civilization over the course of its history transforms 1 percent of that material into technological artifacts. This may seem a lot. However, on Earth we are used to an "exponential" growth in the amount of material we can process—a doubling in a certain period, then doubling again and again. If this kind of increase were to continue once an alien civilization became space faring, says Arkhipov, the civilization might plausibly process 1 percent of its asteroidal material into the extraterrestrial equivalent of consumer goods in just a few million years.

Now for the numbers. One percent of 1.8 trillion trillion grams is 18 billion trillion grams. If this amount of asteroidal material is converted into objects of, say, mass 100 grams—the size of a pepper shaker—this comes to 180 billion billion artifacts.

Not every star has a family of planets, although it appears from recent observations of nearby stars that as many as 30 percent do.* And not every planetary system is likely to spawn a space-faring civilization that makes interstellar artifacts. But say, just for the sake of argument, 1 percent of civilizations do. How common will alien artifacts then be?

Inserting the figures, the answer is that every cube of space with sides 130 million kilometers long should have one artifact floating in it. This is a truly enormous volume—roughly the distance of Earth from the Sun. Searching for an alien artifact the size of a pepper shaker in this much empty space makes hunting down the proverbial needle in a haystack child's play. From the cosmic perspective, however, alien artifacts would be surprisingly common.

Imagine, then, that the whole of space, stretching out to the most distant stars, is divided into cubes 130 million kilometers on a side and that, drifting somewhere in each cube, is one alien artifact. Now, the Sun is not sitting still. It is flying through space with Earth and planets in tow. As it does so, it is therefore flying through a cloud of alien artifacts. It's rather like someone running through a cloud of mosquitoes— except that the cloud of alien artifacts is considerably more rarefied than any conceivable cloud of mosquitoes. The question Arkhipov asks is: how often will Earth run into an alien artifact?

Crudely speaking, any artifact that happens to fly through the space between Earth and the Sun will stand a chance of running into Earth. Think of the space inside Earth's orbit as a circular target flying through a swarm of mosquitoes. Actually, the target can appear bigger than Earth's orbit because the gravity of the Sun can pull in passing artifacts even if they pass farther out from the Sun than Earth. It all depends on their speed. If an artifact races through the solar system at high velocity, it will have to come within Earth's orbit in order to hit Earth. If, however, it drifts slowly through the solar system, it can be snared from farther afield.

How fast will these artifacts be moving? Well, they come from the stars, and the Sun is moving relative to the nearby stars, and these stars in turn have a spread of velocities. Arkhipov calculates that because of this motion, we can expect a piece of space garbage to come at us at about 32 kilometers per second on average, or 115,000 kilometers per hour.

---

* The fraction with planets whose existence we have deduced is nearer one in ten. However, many more show evidence of disks of matter swirling around them, which may be embryonic planetary systems.

At this point, we have almost everything we need in order to calculate how often Earth runs into an alien artifact. We know the size of the target that the space inside Earth's orbit presents to mosquito-like artifacts. We have an estimate of how common such artifacts are throughout space. And we know how fast the Earth-Sun target is flying through the cloud of artifacts.

As the target travels through space, it "sweeps out" a cylindrical volume of space. Any artifact that happens to be in that cylinder can potentially fall to Earth. To fall to Earth, it must actually run smack bang into Earth, and Earth presents an even smaller target than the space within Earth's orbit. However, taking this into consideration as well, Arkhipov finally answers the question: how many hundred-gram alien artifacts have fallen to Earth in its 4.6 billion-year history? The answer, incredible as it sounds, is four thousand.

It's an amazing figure. But it applies only if the underlying assumptions are correct. In other words, if 1 percent of the planetary system spawns a space-faring civilization that turns an amount of material equivalent to 1 percent of the asteroid belt into hundred-gram artifacts.

The numbers are, of course, flexible. For instance, it could be that only one in ten thousand planetary systems produces a space-faring civilization during its lifetime. This would mean revising down the number of alien artifacts on Earth to a mere forty. Not a lot, perhaps. But emphatically not zero. Or say space-faring civilizations transform a smaller mass of material into artifacts or transform the mass into bigger artifacts. Well, in both cases, the figure of four thousand will need to be scaled down. But the amazing thing is that it has to be scaled down by a factor of more than four thousand to be less than one.

Arkhipov's extraordinary conclusion is that if a reasonable proportion of planetary systems produces space-faring extraterrestrials, then alien garbage must exist on Earth. This assumes that such material does not burn up completely as it plunges down through the atmosphere. However, Arkhipov believes that moderate-sized artifacts have a good chance of surviving this ordeal by fire, at least in part.

The evidence of extraterrestrials, according to Arkhipov, could literally be beneath our feet. Consequently, scientists should seriously consider looking for alien artifacts in geological strata and among unusual meteorites. "The discovery of ET evidence seems possible not only in the sky but on Earth as well!" he says.

Comparisons can be made with SETI, the search for extraterrestrial intelligence. Whereas our radio dishes have presented a target for alien radio traffic for little more than forty years, Earth, and especially the Moon, have been sitting targets for alien garbage for more than four billion years. That's one hundred million times longer. Moreover, a basic assumption of SETI is that extraterrestrials not only possess a desire to communicate but that they use a means—radio or optical signals—that we at our present stage of advancement are able to recognize. By contrast, the mere existence of extraterrestrials is enough to guarantee that their garbage eventually falls onto Earth and the Moon. "The search for alien space debris is the missed chance of modern science," says Arkhipov.

Finding Alien Artifacts

Probably the best place to look for alien artifacts is on the Moon, as Arthur Clarke guessed. The Moon, after all, has not been weathered or remade by geological forces apart from meteorite impacts. "The lunar surface must be studied by archaeologists," says Arkhipov.

The Moon, however, is beyond our reach at present and our best bet is our own planet. The first thing to say is that almost two-thirds of Earth is ocean. It follows that this is the most likely place for an alien artifact to come down. The pressure at the foot of the deepest ocean trenches is cripplingly high and we cannot go there, but must send our robotic emissaries instead. As oceanographers are fond of saying, the bed of the ocean is less well known than the surface of the Moon. It does not bode too well for finding the odd alien artifact the size of a pepper shaker.

So much for the oceans. What of Earth's dry land? This must contend with the remorseless effect of wind, rain, and ice, which over time can weather away even the tallest mountains. But even these forces pale into insignificance beside the geological ones that, over hundreds of millions of millions of years, have seen new oceans open up and continents dive into oblivion in the magma beneath our feet. The prospects for an alien artifact falling to Earth do not look good. In fact, any artifact that fell to Earth more than a billion years ago has probably long ago been dragged down into Earth, crushed, and transformed by the heat and pressure of the planet's interior.

However, it may be that few alien artifacts fell to Earth in the first few billion years of its history. There is a hidden assumption in Arkhipov's reasoning and that is that space-faring civilizations have always been around. It turns out, however, that the heavy atoms necessary for life, such as carbon and oxygen and iron, are baked inside the ovens of stars before being spewed into space to be incorporated into new stars. This process ensures that successive generations of stars have been richer in heavy elements and it may be that a certain threshold level of heavy atoms is needed before life-bearing planets like Earth are possible. Consequently, Earth could be one of the first, and intelligent life might not have arisen elsewhere long before it did on Earth.

For a number of reasons, therefore, any artifacts we might find would be likely to be from the last billion years. How then would we recognize one? It stands to reason that we are unlikely to find an alien transistor radio in rocks containing dinosaur bones.

Here we have a major problem. Would people from the nineteenth century recognize a silicon chip? Would they realize that it was an artifact of an advanced technological civilization? Would they realize that it could carry out millions, or even billions, of calculations a second? Nineteenth-century chemists might conclude that the chip was made of an element called silicon and that there were traces of other elements such as gold. But they would be unlikely to guess its purpose. The situation would be even worse if the chip had lain around for years so that it was weathered and eroded.

And here we are talking of a span of only a hundred years. An advanced extraterrestrial civilization might be thousands, even millions, of years ahead of us. Its artifacts might be as unrecognizable to us as a dishwasher is to an ant, or even an amoeba. In the words of Arthur C. Clarke: "Any technology that is sufficiently advanced is indistinguishable from magic."

Our only hope, it would seem, would be to find a puzzling piece of rock or metal with an unusual chemical composition or even an unusual nuclear composition.

Somewhere in the world a puzzling artifact is lying in a museum. Perhaps nobody has noticed it for a century or more. Or perhaps, at this very moment, a curator is taking it out of a glass case, looking at it, and scratching his or her head in bafflement. Will the curator take it to be chemically analyzed? Or will the curator put it back in the case and forget about it forever? We can only hope that doesn't happen.

Glossary

absolute zero. The lowest temperature attainable. When an object is cooled, its atoms move more and more sluggishly. Absolute zero, which is equivalent to -273.15 on the Celsius scale, is considered to be the temperature at which they stop moving altogether. (This is not entirely true because the Heisenberg uncertainty principle produces a residual jitter even at absolute zero.)

accretion disk. A CD-shaped disk of in-swirling matter that forms around a strong source of gravity such as a black hole. Because gravity weakens with distance from its source, the gas and dust in the outer regions of the disk orbit more slowly than those in the inner regions. Friction between regions where matter is traveling at different speeds can heat the disk to millions of degrees. Quasars are thought to owe their prodigious brightness to ferociously hot accretion disks surrounding giant black holes.

amino acid. A type of molecule used by living things as the basic component of proteins.

anthropic principle. The idea that the universe is the way it is because, if it were not, we would not be here to notice it. In other words, our existence is an important scientific observation. This principle can be used to rule out all laws of physics that do not lead to the emergence of stars, planets, and life.

antimatter. A large accumulation of antiparticles. Antiprotons, antineutrons, and positrons can come together to make antiatoms; there is nothing in principle to rule out the possibility of antistars, antiplanets, and antilife. One of the greatest mysteries of physics is why we appear to live in a universe made exclusively of matter when the laws of physics seem to predict a fifty-fifty mix of matter and antimatter.

antiparticle. A particle associated with another subatomic particle with opposite properties such as electrical charge. For instance, the negatively charged electron is twinned with a positively charged antiparticle known as a positron. When a particle and its antiparticle meet, they self-destruct, or "annihilate," in a flash of high-energy light, or gamma rays.

arrow of time. See "thermodynamic arrow of time."

asteroid. A small rocky body that orbits a star. Such bodies are probably the "builders' rubble" left over from the formation of planets.

asteroid belt. The swath of asteroids that orbits the Sun between the orbits of Mars and Jupiter. The largest asteroid, Ceres, is about nine hundred kilometers in diameter.

atom. The building block of all normal matter. An atom consists of a nucleus orbited by a cloud of electrons. The positive charge of the nucleus is exactly balanced by the negative charge of the electrons. An atom is about one ten-millionth of a millimeter across.

atomic energy. See "nuclear energy."

atomic nucleus. The tight cluster of protons and neutrons (a single proton in the case of hydrogen) at the center of an atom. The nucleus contains more than 99.9 percent of the mass of an atom.

bacterium. A single-celled organism, among the simplest known living organisms.

Big Bang. The titanic explosion in which the universe is thought to have been born between twelve and fourteen billion years ago.

Big Bang theory. The idea that the universe began in a superdense, superhot state twelve to fourteen billion years ago and has been expanding and cooling ever since.

Big Crunch. The implosion in which the universe might end. If there is enough matter in the universe, its gravity will one day halt and reverse the universe's expansion so that it shrinks down to a dense ball; a sort of mirror-image of the Big Bang.

black hole. A warped space-time left behind when a massive body has been crushed out of existence by its own gravity. Nothing, not even light, can escape the vicinity of a black hole. The universe appears to contain two distinct types of black hole—stellar-sized black holes, formed when massive stars can no longer generate enough internal heat to counterbalance the gravity trying to crush them, and "supermassive" black holes. Most galaxies appear to have supermassive black holes in their centers. They range from millions of times the mass of the Sun in our Milky Way to billions of solar masses in the powerful quasars. It is also possible that our universe contains tiny black holes, relics of the Big Bang.

causality. The idea that an effect is always preceded by a cause. For instance, you get wet from the rain after it starts raining and not before. Causality is a much-cherished principle in physics.

causality violation. The idea of a cause being preceded by an effect—for example, getting wet from rain before it starts raining, or dying before you are born. Causality violation generally gives physicists palpitations.

CERN. The European laboratory for particle physics near Geneva, Switzerland.

classical physics. Nonquantum physics. In effect, all physics before 1900, when the German physicist Max Planck proposed that energy might come in discrete chunks, or "quanta." Einstein was the first to realize that this idea was totally incompatible with all physics that had gone before.

closed timelike curve. A region of space-time so dramatically warped that time loops back on itself in much the same way that space loops back on itself on an athletics track. A CTC, in common parlance, is a time machine.

comet. A small icy body—usually only a few kilometers across—that orbits a star. Most comets orbit the Sun beyond the outermost planets in an enormous cloud known as the Oort cloud. Like asteroids, comets are builders' rubble left over from the formation of the planets.

conservation law. A law of physics that states that a quantity can never change. For instance, the law of the conservation of energy states that energy can never be created or destroyed, only converted from one form to another; the chemical energy of gasoline can be converted into the energy of motion of a car.

Copenhagen interpretation. For many years, the standard explanation of why the microscopic, or "quantum," world appears so different from the everyday world. According to the Copenhagen interpretation, a quantum object such as an atom can be in several places at once, but it is the act of being observed that forces it to plump for one place and one place only. Because what constitutes an "observation" is ill defined, the Copenhagen interpretation is itself open to interpretation.

Copernican principle. The idea there is nothing special about our position in the universe, either in space or in time. This is a generalized version of Copernicus's recognition that Earth is not in a special position at the center of the solar system but is just another planet circling the Sun.

cosmic rays. High-speed atomic nuclei, mostly protons, from space. Low-energy ones come from the Sun; high-energy ones probably come from supernovae. The origin of ultra-high-energy cosmic rays, particles millions of times more energetic than anything we can currently produce on Earth, is one of the great unsolved puzzles of astronomy.

cosmology. The science whose subject matter is the origin, evolution, and fate of the entire universe.

cosmos. Another word for universe.

dark matter. Matter in the universe that gives out no light. Astronomers know it exists because the gravity of the invisible stuff bends the paths of visible stars and galaxies as they fly through space. There is at least ten times as much dark matter in the universe as visible matter. The identity of the dark matter is the outstanding problem of astronomy.

decoherence. The mechanism that destroys the weird quantum nature of a body—so that, for instance, it appears localized rather than in many different places simultaneously. Decoherence occurs if the outside world gets to know about the body. The knowledge may be taken away by a single photon of light or an air molecule that bounces off the body. Because big bodies such as tables are continually struck by photons and air molecules, and cannot remain isolated from their surroundings for long, they lose their ability to be in many places at once in a fantastically short time—far too short for us to notice.

density. The mass of an object divided by its volume. Air has a low density and iron has a high density.

deuterium. A rare isotope of hydrogen. Deuterium contains a neutron as well as a proton in its nucleus.

dimension. An independent direction in space-time. The familiar world around us has three spatial dimensions (left-right, forward-backward, and up-down) and one temporal dimension (past-future). Superstring theory requires the universe to have six extra spatial dimensions. These differ radically from the other dimensions because they are rolled up small.

DNA. Deoxyribonucleic acid, the ultimate store of molecular information for all cells.

double-slit experiment. An experiment in which particles are shot at a screen with two closely spaced, parallel slits cut in it. On the far side of the screen, the particles mingle, or "interfere," with one another to produce a characteristic "interference pattern" on a second screen. The pattern forms even if the particles are shot at the slits one at a time, with long gaps between shots—in other words, when there is no possibility of their mingling. This result, claimed the American physicist Richard Feynman, highlighted the "central mystery" of quantum theory.

electromagnetic force. One of the four fundamental forces of nature.

electron. A negatively charged subatomic particle typically found orbiting the nucleus of atoms. Most people believe it is a truly elementary particle, incapable of being subdivided.

electron bubble. A bubble formed when an electron lodged in liquid helium repels all the surrounding atoms.

energy. The capacity to perform work; it is a quantity that can never be created or destroyed, only converted from one form to another. (Energy is almost impossible to define. Among the many familiar forms of energy are heat energy, energy of motion, electrical energy, and sound energy.)

energy, conservation of. The principle that energy can never be created or destroyed, only converted from one form to another.

entropy. A measure of the disorder of a body. More precisely, it is the number of possible ways of rearranging the component parts of a body while leaving its overall appearance unchanged. According to the second law of thermodynamics, one of the cornerstones of physics, entropy can never decrease. Though it may not appear to be, this statement entirely equivalent to the statement that heat cannot pass from a cold body to a hot body.

escape velocity. The speed a body must attain in order to escape forever from the gravitational grip of another body. The escape velocity from Earth's surface is about 11 kilometers per second, and from the Sun about 618 kilometres per second.

Europa. One of Jupiter's giant moons, which are also known as Galilean moons. Europa is of particular interest because of the strong suspicion that, beneath its icy crust, there lies a giant ocean—the biggest ocean in the solar system.

event horizon. The one-way "membrane" that surrounds a black hole. Anything that falls through—whether matter or light—can never get out again.

expanding universe. The fleeing of the galaxies from each other in the aftermath of the Big Bang.

extrasolar planets. Planets orbiting stars other than the Sun.

extremeophiles. Living organisms that can survive extreme conditions such as intense cold, boiling temperatures, or total darkness.

fine-tuning of the laws of physics. The observation that the laws of physics are just right to permit the existence of stars, planets, and life. If the force of gravity, for instance, were even a few percent weaker or stronger than we find it, human beings would never have developed.

force-carrying particle. A microscopic conveyor of a force. Forces arise through the continual exchange of such particles, in much the same way that the continual exchange of a tennis ball between tennis players results in a small force between them.

formal system. A system that is the basic building block of mathematics. It consists of a set of self-evident truths, or "axioms," and the conclusions, or "theorems," that can be deduced from them.

fundamental force. One of the four basic forces that are believed to underlie all phenomena. The four forces are gravitational force, electromagnetic force, strong force, and weak force. The strong suspicion among physicists is that these forces are merely facets of a single superforce. Experiments have already shown the electromagnetic and weak forces to be different aspects of a single force, dubbed the electroweak force. Theorists have also devised theories—known as Grand Unified Theories, or GUTs—in which the electroweak and strong forces appear as different faces of a single coin. As yet, however, no single GUT theory has been confirmed by experiments.

fundamental particle. A particle that is one of the basic building blocks of all matter. Currently, physicists believe there are six different quarks and six different leptons, making a total of twelve truly fundamental particles. The hope is that the quarks will turn out to be merely different faces of the leptons.

galaxy. A great island of stars, one of the components of the universe. Our own island, the Milky Way, is spiral in shape and contains several hundred thousand million stars.

general theory of relativity. Einstein's theory of gravity, which shows gravity to be nothing more than the warpage of space-time. The theory incorporates several ideas that were not incorporated in Newton's theory of gravity. One is that nothing, not even gravity, can travel faster than light. Another is that all forms of energy have mass, and so are sources of gravity. Among other things, the theory predicted black holes, the expansion of the universe, and that gravity would bend the path of light.

gluon. A force-carrying particle of the strong force.

gravitational force. The weakest of the four fundamental forces of nature. Gravity is approximately described by Newton's universal law of gravity, but more accurately described by Einstein's theory of gravity — the general theory of relativity. General relativity breaks down at the singularity at the heart of a black hole and the singularity at the birth of the universe. Physicists are currently looking for a better description of gravity. That theory, already dubbed quantum gravity, will explain gravity in terms of the exchange of particles called gravitons.

gravitational lensing. The magnification of the light of a distant astronomical object by another object between the distant object and Earth. The intervening object, which could be a galaxy or a star, is known as a gravitational lens. The effect occurs because gravity can bend the trajectory of light. This means that light, as it passes the lensing body on its way to Earth, is bent toward the body, which in effect acts like a converging glass lens.

gravitational slingshot. The boost in speed a body such as a space probe or an asteroid gets from the gravity of a massive body such as a planet. The maximum effect occurs if the body flies close to the planet.

graviton. A hypothetical force-carrying particle of the gravitational force.

gravity. See "gravitational force."

greenhouse gas. A gas that lets through visible light but absorbs heat radiation, or infrared light. Such gases in Earth's atmosphere act like a blanket, trapping warmth near the surface. The most effective greenhouse gas is water vapor, without which Earth's average temperature would be many tens of degrees below freezing.

half-life. The time it takes half the atoms in a radioactive sample to disintegrate. After one half-life, half the atoms will be left; after two half-lives, a quarter; after three, an eighth, and so on. Half-lives can vary from the merest split second to many billions of years.

Heisenberg uncertainty principle. A principle of quantum theory that states that there are pairs of quantities, such as a particle's location and speed, that cannot simultaneously be known with absolute precision. The uncertainty principle puts a limit on how well the product of such a pair of quantities can be known. In practice, this means that if the speed of a particle is known precisely, it is impossible to have any idea where the particle is. Conversely, if the location is known with certainty, the particle's speed is unknown. By limiting what we can know, the Heisenberg uncertainty principle imposes a "fuzziness" on nature. If we look too closely, everything blurs like a newspaper photograph dissolving into meaningless dots.

helium. The second-lightest element in nature, and the only one to have been discovered on the Sun before it was discovered on Earth. Helium is the second-most-common element in the universe after hydrogen, accounting for about 10 percent of all atoms. Most helium was forged in the Big Bang. Below 4.2 degrees above absolute zero, helium condenses into a liquid. Below 2.17 degrees, it becomes a superfluid with ability to run uphill and squeeze through impossibly small holes.

horizon. The boundary of the observable universe. The horizon is much like the horizon that surrounds a ship at sea. The reason for the universe's horizon is that light has a finite speed and the universe has been in existence for only a finite time. This means that we see only objects whose light has had time to reach us since the Big Bang. The observable universe is therefore like a bubble centered on Earth, with the horizon being the surface of the bubble. Every day the universe gets older (by one day), so every day the horizon expands outward and new things become visible, just as ships are seen coming over the horizon at sea.

horizon problem. The problem that far-flung parts of the universe that could never have been in contact with one another, even in the Big Bang, have almost identical properties such as density and temperature. Technically, they have always been beyond one another's horizon. The theory of inflation provides a way for such regions to have been in contact during the Big Bang and so can potentially solve the horizon problem.

hydrogen. The lightest element in nature. A hydrogen atom consists of a single proton orbited by a single electron. Close to 90 percent of all atoms in the universe are hydrogen atoms.

inflation, theory of. The idea that in the first split second of the creation, the universe underwent a fantastically fast expansion. In a sense, inflation preceded the conventional Big Bang explosion. If the Big Bang is likened to the explosion of a grenade, inflation is like the explosion of an H-bomb. Inflation can solve some problems with the Big Bang theory, such as the horizon problem.

infrared. A type of invisible light given out by warm bodies. Infrared light has a longer wavelength than visible light.

interstellar dust. Tiny grains of dust that hang in the space between stars. They are believed to have been formed in the atmospheres of dying stars and ejected into space. There is still some debate about exactly what they are made of.

interstellar medium. A tenuous medium consisting of gas and dust floating between the stars. In the vicinity of the Sun, this gas comprises about one hydrogen atom in every three cubic centimeters, making it a vacuum far better than anything achievable on Earth.

interstellar planet. A hypothetical planet that wanders alone in the deep freeze of interstellar space. Such planets could have been ejected from the vicinity of stars during the process of planet formation.

interstellar space. The space between the stars.

isotope. One possible form of an element. Isotopes are distinguishable by their differing masses. For instance, chlorine comes in two stable isotopes, with masses of 35 and 37. The mass difference is the result of a differing number of neutrons in their nuclei. Chlorine-35 contains eighteen neutrons and chlorine-37 contains twenty neutrons. (Each contains the same number of protons—seventeen—because the number of protons determines the identity of an element.)

Jupiter. The largest planet in our solar system. It is more massive than all the other planets put together.

Kaluza-Klein particle. A new particle that appears in Kaluza-Klein theory. In effect, it is an "echo" in the theory's curled-up extra dimensions.

Kaluza-Klein theory. A theory with extra, curled-up dimensions.

lepton. One of a group of subatomic particles including the electron and neutrino. Leptons, along with quarks, are currently thought to be the ultimate building blocks of nature. There are six different quarks and six leptons.

LHC. The Large Hadron Collider, a giant particle accelerator being built at CERN and due for completion in 2006.

light bending. See "gravitational lensing."

light, speed of. The cosmic speed limit—three hundred thousand kilometers per second.

light-year. A convenient unit for expressing distance in the universe. It is the distance light travels in one year, which turns out to be 9.46 trillion kilometers.

luminosity. The total amount of light pumped into space by a celestial body such as a star.

Many Worlds idea. The idea that quantum theory describes everything, not simply the microscopic world of atoms and their constituents. Because quantum theory permits an atom to be in two places at once, this implies that a large thing such as a table can also be in two places at once. According to the Many Worlds idea, however, the mind of the person observing the table splits into two—one that perceives the table in one place and another that perceives it in another. There remains the sticky problem of explaining why, if atoms can be in two or more places at once, someone's mind cannot be in two states at once, simultaneously perceiving the table in two places at once. This problem is solved by the phenomenon of decoherence.

mass. A measure of the amount of matter in a body. Mass is the most concentrated form of energy. A single gram contains the same amount of energy as would be released by the explosion of one hundred tons of dynamite.

meteorite. A chunk of interplanetary rubble that falls to Earth.

Milky Way. Our galaxy.

mirror matter. Matter composed of mirror particles.

mirror particles. Fundamental particles that interact in a way that is the mirror image of the way in which normal particles interact. If mirror matter existed, every particle would have its mirror counterpart.

mirror universe. If mirror matter existed, there could be a mirror universe occupying the same space as our universe. It could contain mirror planets, mirror stars, and mirror galaxies—and all would essentially be invisible.

molecule. A collection of atoms held together by electromagnetic forces. One atom, carbon, can link with itself and other atoms to make a huge number of molecules. For this reason, chemists divide molecules into "organic" (those based on carbon) and "inorganic" (the rest).

molecule, interstellar. One of the more than one hundred kinds of molecule found floating in space. Interstellar molecules include ethyl alcohol and the simple amino acid known as glycine. Each has an optical "fingerprint"—a characteristic set of wavelengths at which it emits or absorbs light. By this means, astronomers can identify a molecule when they pick up its light with their telescopes.

multiverse. The hypothetical enlargement of the cosmos in which our universe turns out to be one among an enormous number of separate and distinct universes. Most universes are dead and uninteresting. Only in a tiny subset do the laws of physics promote the emergence of stars and planets and life.

natural selection of universes. A variant of the idea of the self-reproducing universe. Instead of new universes being spawned automatically—perhaps inside black holes—intelligent life takes over the universe-building business. This means that universes in which the laws of physics promote the emergence of intelligent life reproduce at the expense of other universes. The eventual result is a multiverse in which most universes contain life.

nebula. A cloud of tenuous gas in space. If young, hot stars are embedded in the gas, they will cause it to glow brightly. If there are no such stars, the nebula may still reveal itself as a black splotch that blots out the light of more distant stars.

nemesis. A hypothetical companion star of the Sun, which was proposed to explain the regular recurrence of mass extinctions in Earth's history. The regularity is much disputed. Such a companion, to have escaped detection, would have to be extremely dim and in a highly unusual orbit.

neuron. A cell in the nervous system that receives and transmits information in the form of electrical pulses. In the brain, such cells underlie the processes of thought.

neutrino. A neutral subatomic particle with a very small mass that travels close to the speed of light. Neutrinos hardly ever interact with matter. However, when created in huge numbers they can blow a star apart, as in a supernova.

neutron. One of the two main components of the atomic nucleus at the center of atoms. Neutrons have essentially the same mass as protons but carry no electrical charge. They are unstable outside a nucleus and disintegrate in about ten minutes.

Newton's universal law of gravity. The idea that all bodies pull on each other across space with a force that depends on the product of their individual masses and the inverse square of their distance apart. In other words, if the distance between the bodies is doubled, the force becomes four time weaker; if it is tripled, nine times weaker; and so on. Newton's theory of gravity is perfectly good for everyday applications but turns out to be an approximation. Einstein provided an improvement in the form of the general theory of relativity.

nonlocality. The spooky ability of objects subject to quantum theory to continue to "know about" one another's states even when separated by a large distance.

nuclear energy. The excess energy released when one atomic nucleus changes into another atomic nucleus.

nucleon. A proton or neutron, the two building blocks of the atomic nucleus.

nucleus. See "atomic nucleus."

Oort Cloud. A cloud of comets thought to orbit the Sun beyond the orbit of the outermost planet. Estimates put the total number of comets in the Oort Cloud at up to one hundred billion.

ortho-positronium. A form of positronium in which the electron and positron spin in the same direction.

panspermia. The idea that the seeds of life spread across space from planetary system to planetary system, and that simple life on Earth was therefore "seeded" from the stars.

para-positronium. A form of positronium in which the electron and positron spin in opposite directions.

particle accelerator. A giant machine, often in the shape of a racetrack, in which subatomic particles are accelerated to high speeds and smashed into one another. In such collisions, the energy of motion of the particles becomes available to create new particles, which are what interest particle physicists.

particle physicist. A researcher who engages in particle physics.

particle physics. The quest to discover the fundamental components and the fundamental forces of nature.

photon. A particle of light.

physics, laws of. The fundamental laws that orchestrate the behavior of the universe.

Planck energy. The super-high energy at which gravity becomes comparable in strength to the other fundamental forces of nature.

Planck length. The fantastically tiny length scale at which gravity becomes comparable in strength to the other fundamental forces of nature. The Planck length is a trillion trillion times smaller than an atom. It corresponds to the Planck energy. Small distances are synonymous with high energies because of the wave nature of matter. Because of this nature, when a particle is confined to a small volume, its wave is scrunched up. Rather than being spread out, it becomes a localized, choppy wave. In other words, it becomes more violent and the energy of the wave increases.

planet. A small sphere-shaped body orbiting a star. A planet does not produce its own light but shines by reflecting the light of its star.

polymer. A large daisy-chain-like molecule formed from repeating a basic arrangement of atoms.

positron. The antiparticle of the electron.

positronium. An electron and a positron in orbit around each other.

prebiology. The idea that the earliest steps in the emergence of life occurred somewhere other than on Earth and so preceded biological evolution on Earth.

proton. One of the two main building blocks of the nucleus. Protons carry a positive electrical charge, equal and opposite to that of electrons.

protoplanetary nebula. A cloud of interstellar gas and dust surrounding a newborn star from which a system of planets forms.

QED. See "quantum electrodynamics."

quanta. The smallest chunks into which something can be divided. For instance, photons are quanta.

quantum chromodynamics. The theory of the strong force between quarks. The force arises as a consequence of the exchange of particles known as gluons.

quantum computer. A machine that exploits the idea that quantum systems such as atoms can be in many different states at once to carry out many calculations at once. The best quantum computers can manipulate only a handful of binary digits, or bits, but in principle such computers could massively outperform conventional computers.

quantum electrodynamics. The theory of how light, in the form of photons, interacts with matter. This theory explains almost everything about the everyday world, from why the ground beneath our feet is solid to how a laser works, from the chemistry of metabolism to the operation of computers.

quantum superposition. The situation in which a quantum object such as an atom is in more than one state at a time. It might, for instance, be in many places simultaneously. It is the interaction, or "interference," between the individual states in the superposition that is the basis of all quantum weirdness. Decoherence prevents such interaction and therefore destroys quantum behavior.

quantum theory. Essentially, the theory of the microscopic world of atoms and their constituents. Those who favor the Many Worlds interpretation believe it also describes the large-scale world.

quark. A particle from which protons and neutrons of atoms are made. Quarks, along with leptons, are currently thought to be the ultimate building blocks of nature. There are six different quarks and six leptons.

quark-hadron phase transition. The time, a mere millionth of a second after the moment of creation, when the universe was cool enough for clumps of quarks to condense into protons and neutrons. Protons and neutrons are actually "bags" of quarks; one combination of three quarks makes up the proton and another combination the neutron.

quasar. A galaxy that derives most of its energy from matter heated to millions of degrees as it swirls into a central, giant black hole. Quasars can generate as much light as a hundred normal galaxies from a volume smaller than the solar system, making them the most powerful objects in the universe.

radioactive decay. The disintegration of heavy atoms, which are unstable, into lighter atoms, which are stable.

radioactivity. The property of atoms that undergo radioactive decay.

Schrödinger equation. An equation that governs the way in which the wave function describing, say, an atom changes with time. The Schrödinger equation therefore makes it possible to predict the atom's future behavior.

search for extraterrestrial intelligence. A search that uses telescopes to scan the sky for either radio or optical signals from extraterrestrial civilizations.

self-reproducing universe. A universe that gives birth to other universes.

SETI. See "search for extraterrestrial intelligence."

singularity. The location where the fabric of space-time ruptures and therefore cannot be understood by Einstein's theory of gravity, the general theory of relativity. There was a singularity at the beginning of the universe. There is also one in the center of every black hole.

solar system. The Sun and its family of planets, moons, comets, and other assorted rubble.

space junk. The dead satellites, discarded rocket casings, and so forth that girdle the Earth, creating a hazard for space vehicles.

space-time. A single entity composed of space and time, which, in the general theory of relativity, are seen to be essentially the same. Gravity is the warpage of space-time.

spectral line. Light emitted or absorbed by atoms and molecules. If they swallow more light than they emit, the result is a dark line in the spectrum of a celestial object. Conversely, if they emit more than they swallow, the result is a bright line.

spectroscopy. The technique of measuring the spectrum of an object.

spectrum. The rainbow of colors created when light is separated into its constituent parts.

star. A giant ball of gas that replenishes the heat it loses to space by means of nuclear energy generated in its core.

string theory. See "superstring theory."

strong force. The powerful short-range force that holds protons and neutrons together in an atomic nucleus.

strong nuclear force. See "strong force."

subatomic particle. A particle smaller than an atom, such as an electron or a neutron.

Sun. The star nearest Earth.

superfluid. A fluid that, below a critical temperature, develops bizarre properties such as the ability to flow uphill and to squeeze through impossibly small holes. The best example is liquid helium, which becomes a superfluid below a temperature of 2.17 degrees above absolute zero. Superfluids owe their weirdness to quantum theory, and are also called "quantum liquids."

superforce. A hypothetical force that unites, or "unifies," the four fundamental forces. The hope is that they will turn out to be mere facets of this superforce.

supernova. A cataclysmic explosion of a star. A supernova may, for a short time, out shine an entire galaxy of a hundred billion ordinary stars. It is thought to leave behind a highly compressed neutron star.

supernova 1987A. A supernova discovered in Large Magellanic Cloud, a satellite galaxy of our galaxy, on 24 February 1987. It was the first supernova to be spotted in our neighborhood for 387 years.

superstring theory. A theory that postulates that the fundamental ingredients of the universe are tiny strings of matter. The strings vibrate in a space-time of ten dimensions. The great payoff of this idea is that it may be able to unite, or unify, quantum theory and the general theory of relativity.

symmetry. A quality of an object that does not change when the object is transformed in some way. For instance, a face that looks the same when it is reflected in a mirror is said to show "mirror symmetry."

temperature. The degree of "hotness" of a body. Temperature is related to the energy of motion of the particles that compose it.

thermodynamic arrow of time. The direction of time associated with entropy increasing, people growing old, mugs breaking, and the like.

thermodynamics, second law of. The decree that entropy cannot ever decrease. This is equivalent to saying that heat can never flow from a cold body to a hot body.

time dilation. The slowing down of time for an observer moving close to the speed of light or experiencing strong gravity.

time loop. See "closed timelike curve."

time machine. See "closed timelike curve."

time travel. Travel into the past or future (at a rate of more than one year per year, which we all do).

time-travel paradox. A nonsensical situation that time travel appears to permit. The most famous is the "grandfather paradox," in which someone goes back in time and shoots his or her grandfather. How, then, could the shooter have been born to go back in time and commit the act?

Tunguska event. A giant explosion that occurred on 30 June 1908 in the basin of the Tunguska River in Siberia. Several thousand square kilometers of forest were flattened by a blast equivalent to a ten-megaton H-bomb at an altitude of nine kilometers.

ultimate ensemble theory. The idea that every conceivable set of laws of physics is expressed in a universe somewhere. The idea requires the existence not only of a multiverse, but of the biggest possible multiverse.

ultraviolet. A type of invisible light given out by very hot bodies and responsible for sunburn. Ultraviolet light has a shorter wavelength than visible light.

unification. The idea that at extremely high energy, the four fundamental forces of nature were once one, united in a single theoretical framework.

universe. All there is. This is a flexible term, which once was used for what we now call the solar system. Later, it was used for what we call the Milky Way. Now it is used for the sum total of all the galaxies, of which there appear to be about ten billion within the observable universe.

universe, expansion of. The fleeing of the galaxies from each other in the aftermath of the Big Bang.

universe, observable. All we can see out to the universe's horizon.

vector boson. A particle that is exchanged between particles and whose exchange gives rise to the weak nuclear force.

virtual particle. A particle that has a fleeting existence, popping into being and popping out again according to the constraint imposed by the Heisenberg uncertainty principle.

wave function. A mathematical entity that contains all that is knowable about a quantum object such as an atom. The wave function changes over time according to the Schrödinger equation.

wavelength. The distance it takes for a wave to go through a complete oscillation cycle.

wave-particle nature of matter. The quality of a subatomic particle behaving as a localized billiard-ball-like particle or a spread-out wave.

weak nuclear force. The second force experienced by protons and neutrons in an atomic nucleus, the other being the strong nuclear force. The weak nuclear force can convert a neutron into a proton.

wormhole. A tunnel through space-time that connects widely spaced regions and provides a shortcut.

X *rays.* Rays that make up a high-energy form of light.

Z-*boson.* One kind of vector boson.

Further Reading

Chapter 1

Amis, Martin. *Time's Arrow*. London: Penguin, 1991.

Heinlein, Robert. *The Door into Summer*. New York: Ballantine Books, 1956.

Lightman, Alan. *Einstein's Dreams*. London: Sceptre, 1999.

Schulman, L. S. "Opposite Thermodynamic Arrows of Time." *Physical Review Letters* 83 (31 December 1999): 5419.

———. *Time's Arrow and Quantum Measurement*. Cambridge: Cambridge University Press, 1997.

Chapter 2

Baxter, Stephen. *The Time Ships*. London: Voyager/Harper-Collins, 1995.

Brown, Julian. *Minds, Machines, and the Multiverse*. New York: Simon and Schuster, 1999.

Deutsch, David. *The Fabric of Reality*. London: Penguin, 1997.

Pohl, Frederik. *The Coming of the Quantum Cats*. New York: Ballantine, 1996.

Tegmark, Max. "The Interpretation of Quantum Mechanics: Many Worlds or Many Words," available at http://www.hep.upenn.edu/~max/index.html.

Wilson, Robert Charles. "Divided by Infinity." In *Starlight 2*, edited by Patrick Nielsen Hayden. New York: Tor Books, 1998.

Chapter 3

Maris, Humphrey J. "On the Fission of Elementary Particles and Electrons in Liquid Helium." *Journal of Low-Temperature Physics* 120 (1 August 2000): 173.

Chapter 4

Deimer, T., and M. J. Hadley. "Charge and the Topology of Spacetime." *Classical and Quantum Gravity* 16, no. 11 (November 1999): 3567.

Hadley, Mark J. "A Gravitational Explanation for Quantum Mechanics." Ph.D. diss., University of Warwick, England,1997.

———. "The Logic of Quantum Mechanics Derived from Classical General Relativity." *Foundations of Physics* 10, no. 1 (February 1997): 43.

———. "Spin 1/2 in Classical General Relativity." *Classical and Quantum Gravity* 17 (21 October 2000): 4187.

———. "Topology Change and Context Dependence." *International Journal of Theoretical Physics* 38, no. 5 (May 1999): 1481.

———. Website with links to his papers: http://www.warwick.ac.uk/~mapb.

Pais, Abraham. *Inward Bound*. Oxford: Oxford University Press, 1994.

Chapter 5

Abbott, E. *Flatland*. New York: Dover, 1992.

Antoniadis, Ignatios, Nima Arkani-Hamed, Savas Dimopoulos, and Gia Dvali. "New Dimensions at a Millimetre to a Fermi and Superstrings at a TeV." *Physics Letters B* 436 (July 1998): 257; available at http://xxx.lanl.gov/abs/hep-ph/9804398.

Arkani-Hamed, Nima, Savas Dimopoulos, and Gia Dvali. "The Hierarchy Problem and New Dimensions at a Millimetre." *Physics Letters B* 429 (June 1998): 263; available at http://xxx.lanl.gov/abs/hep-ph/9803315.

Heinlein, Robert. "And He Built a Crooked House." Quoted in Pickover, *Surfing through Hyperspace*, viii.

Pickover, Clifford. *Surfing through Hyperspace*. New York: Oxford University Press, 1999.

Randall, Lisa, and Raman Sundrum. "An Alternative to Compactification." *Physical Review Letters* 83 (6 December 1999): 4690.

Chapter 6

Hawkins, M. R. S. "Dark Matter from Quasar Microlensing." *Monthly Notices of the Royal Astronomical Society* 278 (1996): 787.

——. "Gravitational Microlensing, Quasar Variability, and Missing Matter." *Nature* 366 (18 November 1993): 242.
——. *Hunting Down the Universe.* London: Little Brown, 1997.

Chapter 7

Cramer, John. *Twistor.* New York: Avon, 1986.
Foot, Robert, and Sergei Gninenko. "Can the Mirror World Explain the Ortho-positronium Lifetime Puzzle?" *Physics Letters B* 480 (11 May 2000): 171.
Sheffield, Charles. *The Hidden Matter of McAndrew.* New York: Analog, 1992.

Chapter 8

Asimov, Isaac. *The Gods Themselves.* London: Panther, 1974.
Tegmark, Max. "Is 'The Theory of Everything' Merely the Ultimate Ensemble Theory?" *Annals of Physics* 270 (20 November 1998): 1.
——. Website with links to his papers: http://www.hep.upenn.edu/~max/toe.html.

Chapter 9

Asimov, Isaac. *The Gods Themselves.* London: Panther, 1974.
Gunn, James. "Kindergarten." In *Creations*, edited by Isaac Asimov, George Zebrowski, and Martin Greenberg, 105–7. London: Harrap, 1984.
Harrison, Edward. *Cosmology: The Science of the Universe.* Cambridge: Cambridge University Press, 1991.
——. "The Natural Selection of Universes Containing Life." *Quarterly Journal of the Royal Astronomical Society* 35 (summer 1995): 193.
Smolin, Lee. *The Life of the Cosmos.* Oxford: Oxford University Press, 1997.

Chapter 10

Adams, Fred, and Greg Laughlin. "The Frozen Earth." *Icarus* 145 (June 2000): 614.

Gold, Thomas. *The Deep Hot Biosphere.* New York: Copernicus Books, 1999.

Stevenson, David. "Life-Sustaining Planets in Interstellar Space?" *Nature* 400 (1 July 1999): 32; also available at http://www.gps.caltech.edu/faculty/stevenson/.

Chapter 11

Hoyle, Fred. *The Black Cloud.* London: Penguin, 1972.

Hoyle, Fred, and Chandra Wickramasinghe. *Life on Mars: The Case for a Cosmic Heritage.* London: Clinical Press, 1997.

———. *Our Place in the Cosmos.* London: Phoenix, 1996.

Chapter 12

Arkhipov, A. V. "Astroinfect Effect: Revised Model." *Journal of the British Interplanetary Society* 52 (January 1999): 37.

———. "Earth-Moon System as a Collector of Alien Artifacts." *Journal of the British Interplanetary Society* 51 (June 1998): 181.

———. "On the Possibility of Extraterrestrial-Artefact Finds on Earth." *Observatory* 116 (June 1996): 116.

———. Websites with links to his papers: http://www.astrosurf.com/lunascan/arkhipov2.htm; http://www.astrosurf.com/lunascan/arkhipov3.htm.

Clarke, Arthur C. *2001: A Space Odyssey.* New York: New American Library, 1968.

Index

183

About the Author

Marcus Chown has a first-class degree in physics from the University of London and a master's degree in astrophysics from the California Institute of Technology in Pasadena. Currently, he is the cosmology consultant of the British weekly science magazine *New Scientist*.

Chown's first popular-science book, *Afterglow of Creation*, was published to much acclaim in 1994. (The violinist Yehudi Menuhin declared it his favorite book.) In Britain, it was runner-up for the prestigious Rhône-Poulenc science book prize, and the magazine *Focus* liked it enough to buy 180,000 copies for its readers, making it the most-read popular-science book after Stephen Hawking's *Brief History of Time*.

Chown's second popular-science book, *The Magic Furnace*, was published by Oxford in the United States in the spring of 2000. In Japan, it was chosen as one of the Books of the Year by *Asahi Shimbun*, the world's biggest newspaper.

He lives in Worcestershire, England, with his wife, Karen, who is a cancer nurse.